MATHEMATICS AND ITS APPLICATIONS
Series Editor: Professor G. M. Bell
Chelsea College, University of London

GENERALISED FUNCTIONS

This introduction to generalised functions concentrates on basic techniques and applications in a field now widely recognized to be of considerable interest to engineers and applied mathematicians in control, communications systems, network analysis and design. It is intended both as a working manual and reference text in its own right and as a preparation for the study of more advanced and specialised treatises.

The treatment includes many illustrative worked examples and detailed applications to Laplace and Fourier transforms, Fourier series, and linear systems theory.

An especial feature is the author's clear and fully explained account of a subject which is nearly always found mathematically demanding and somewhat mysterious by the beginner. Assuming little more than a standard background in calculus, he develops the theory of the delta-function and its derivatives and shows how its use can clarify much practical analysis. No previous knowledge of functional analysis is required, all the necessary concepts being introduced as and when required in the text.

Readership: Professional electrical engineers, physicists and applied mathematicians, and also suitable for advanced courses and research work in systems theory, control engineering, and network analysis and design.

GENERALISED FUNCTIONS

MATHEMATICS & ITS APPLICATIONS

Series Editor: Professor G. M. Bell
Chelsea College, University of London

The works in this series will survey recent research, and introduce new areas and up-to-date mathematical methods. Undergraduate texts on established topics will stimulate student interest by including present-day applications, and the series can also include selected volumes of lecture notes on important topics which need quick and early publication.

In all these ways it is hoped to render a valuable service to those who learn, teach, develop and use mathematics.

Mathematical Theory of Wave Motion
   G. R. BALDOCK and T. BRIDGEMAN, University of Liverpool
Mathematical Models in Social Management and Life Management Sciences
   DAVID BURGHES and ALISTAIR D. WOOD, Cranfield Institute of Technology
Modern Introduction to Classical Mechanics and Control
   DAVID BURGHES, Cranfield Institute of Technology and
   ANGELA DOWNS, University of Sheffield
Control and Optimal Control
   DAVID BURGHES, Cranfield Institute of Technology and
   ALEX GRAHAM, The Open University, Milton Keynes
Textbook of Dynamics
   FRANK CHORLTON, University of Aston, Birmingham
Vector & Tensor Methods
   FRANK CHORLTON, University of Aston, Birmingham
Advanced Topics in Operational Research
   BRIAN CONNOLLY, Chelsea College, University of London
Lecture Notes on Queueing Systems
   BRIAN CONNOLLY, Chelsea College, University of London
Mathematics for the Biosciences
   G. EASON, C. W. COLES, G. GETTINBY, University of Strathclyde
Handbook of Hypergeometric Integrals: Theory, Applications, Tables, Computer Programs
   HAROLD EXTON, The Polytechnic, Preston
Multiple Hypergeometric Functions
   HAROLD EXTON, The Polytechnic, Preston
Computational Geometry for Design and Manufacture
   IVOR D. FAUX and MICHAEL J. PRATT, Cranfield Institute of Technology
Applied Linear Algebra
   R. J. GOULT, Cranfield Institute of Technology
Matrix Theory and Applications for Engineers and Mathematicians
   ALEXANDER GRAHAM, The Open University, Milton Keynes
Applied Functional Analysis
   D. H. GRIFFEL, University of Bristol
Generalised Functions: Theory, Applications
   R. F. HOSKINS, Cranfield Institute of Technology
Mechanics of Continuous Media
   S. C. HUNTER, University of Sheffield
Game Theory
   A. J. JONES, Royal Holloway College, University of London
Using Computers
   BRIAN MEEK & SIMON FAIRTHORNE, Queen Elizabeth College, University of London
Spectral Theory of Ordinary Differential Operators
   E. MÜLLER-PFEIFFER, Technical High School, Ergurt
Simulation Concepts in Mathematical Modelling
   F. OLIVEIRA-PINTO, Chelsea College, University of London
Environmental Aerodynamics
   R. S. SCORER, Imperial College of Science and Technology, University of London
Applied Statistical Techniques
   K. D. C. STOODLEY, T. LEWIS and C. L. S. STAINTON, University of Bradford
Liquids and Their Properties: A Molecular and Macroscopic Treatise with Applications
   H. N. V. TEMPERLEY, University College of Swansea, University of Wales and
   D. H. TREVENA, University of Wales, Aberystwyth

GENERALISED FUNCTIONS

R. F. HOSKINS, M.Sc.,
Director of Mathematical Studies,
Cranfield Institute of Technology

ELLIS HORWOOD LIMITED
Publishers   Chichester

Halsted Press: a division of
JOHN WILEY & SONS
New York - Chichester - Brisbane - Toronto

First published in 1979 by

ELLIS HORWOOD LIMITED

Market Cross House, Cooper Street, Chichester, West Sussex, PO19 1EB, England

The publisher's colophon is reproduced from James Gillison's drawing of the ancient Market Cross, Chichester

Distributors:

Australia, New Zealand, South-east Asia:
Jacaranda-Wiley Ltd., Jacaranda Press,
JOHN WILEY & SONS INC.,
G.P.O. Box 859, Brisbane, Queensland 40001, Australia.

Canada:
JOHN WILEY & SONS CANADA LIMITED
22 Worcester Road, Rexdale, Ontario, Canada.

Europe, Africa:
JOHN WILEY & SONS LIMITED
Baffins Lane, Chichester, West Sussex, England.

North and South America and the rest of the world:
HALSTED PRESS, a division of
JOHN WILEY & SONS
605 Third Avenue, New York, N.Y. 10016, U.S.A.

©1979 R. F. Hoskins/Ellis Horwood Limited

British Library Cataloguing in Publication Data

Hoskins, R F
  Generalised functions. – (Mathematics & its applications).
  1. Distributions, Theory of (Functional analysis)
  I. Title II. Series
  515'.782   QA324   79-40995

  ISBN 0-85312-105-2 Lib (Ellis Horwood Limited, Publishers)
  ISBN 0-85312-106-0 Pbk (Ellis Horwood Limited, Publishers)
  ISBN 0-470-26608-2   (Halsted Press)

Typeset in Press Roman by Ellis Horwood Limited.
Printed in Great Britain by Unwin Brothers Limited, Woking.

Contents

Author's Preface

The study of generalised functions is now widely recognised to be of considerable interest and importance, both to applied mathematicians and to a large class of engineers working in such fields as control, communications systems, electrical network theory, and so on. The comprehensive treatises currently available, notably those by Jones, Zemanian, or Gelfand and Shilov, are aimed at readers with some considerable mathematical sophistication, and are generally too demanding for the non-specialist. On the other hand, the abbreviated treatments often relegated to the appendices of various engineering texts are necessarily sketchy and are usually of little real help to the beginner.

The object of this book is to provide an elementary but reasonably comprehensive introduction to the theory and applications of generalised functions. It is hoped that it may be used as a working manual in its own right, as well as serving as a preparation for the study of more advanced treatises. The treatment in this book is confined to functions of a single real variable and, as might be expected, is largely devoted to the properties of the delta function and its derivatives. Little more than a standard background in calculus is assumed throughout the greater part of the text, and the attention is focused primarily on techniques with a liberal selection of worked examples and exercises.

Such results from elementary analysis as are needed in the sequel are summarised in Chapter 1, and the study of generalised functions proper begins with the introduction of the delta function itself in Chapter 2. Chapter 3 then develops the properties of the delta function and its derivatives on a heuristic basis, but with due attention to the dangers involved in an uncritical and too free use of the conventional symbolism. The intent is to give the reader a fairly comprehensive tool-kit of rules concerning the manipulation of algebraic expressions involving delta functions and derivatives of delta functions. Chapters 4, 5, and 6 deal with applications. In Chapter 4 we give an introduction to the fundamentals of linear systems theory which is, in some ways, perhaps the most natural and obvious setting for illustrating the significance and use of generalised functions. Certainly it is the case that students often find a discussion of, say, impulse response particularly helpful in dispelling some of the mystery which so often attends first acquaintance with delta functions. Chapter 5 is a more-or-less orthodox account of elementary Laplace transform theory, but with the delta function brought early into the treatment and made use of in the subsequent development. Similarly, Chapter 6 covers standard material in Fourier analysis but again with strong appeals to the delta function both in order to clarify concepts and to simplify techniques.

The last two chapters are different in character. Hitherto the aim has been to make the student familiar with the use of those forms of generalised functions which are likely to occur most frequently in elementary applications. Chapter 7 begins with a detailed and comprehensive discussion of some of the less familiar kinds of generalised functions which usually seem to cause difficulty to the beginner. In the remainder of the chapter there is an elementary account of standard normed linear space theory which leads to an introductory treatment of distribution theory proper, in the sense of Schwartz. It is hoped that this will provide a link between the preceding heuristic work and the more rigorous and mathematically demanding aspects of the subject to be found in specialised and exhaustive texts on distributions. Finally in Chapter 8 we give a comparatively brief and elementary outline of what is essentially the theory of the Lebesgue-Stieltjes integral. This allows the reader to master, one hopes fairly painlessly, the salient features of the Lebesgue theory in general and also to appreciate the significance of the delta function as a measure.

Throughout, the more difficult sections and examples are marked with an asterisk (*) and may be omitted on a first reading. We use the symbol ■ where necessary, to indicate the end of a proof.

This book arose largely from a course of lectures given to a mixed class of control theorists and mathematics students at Cranfield Institute of Technology. My thanks are due to the students, past and present, whose reactions to those lectures helped to shape the book in its final form. I should also like to thank Professor George Bell for his advice and helpful criticism, Ellis Horwood Ltd., for the care and attention lavished on the printing of a complex and exacting text, and last, but not least, my wife for her patience throughout a lengthy period of preparation and revision.

CHAPTER 1

Results from Elementary Analysis

1.1 FUNCTIONS

1.1.1 Unless stated otherwise we always take the term "function" to mean a real, single-valued function of a single real variable (usually understood to be time, t). The range of values of the variable for which the function is defined is called the **domain of definition** of the function. Normally we use single letters such as f, g, h, to denote functions, and understand an expression like $f(t)$ to mean the value which the function f assumes at the point, or instant, t. For example, we may write $f \geqslant g$ to indicate that, for every value of t, the number $f(t)$ is always greater than or equal to the corresponding number $g(t)$:

   $f \geqslant g$ if and only if $f(t) \geqslant g(t)$ for every value of t at which the two functions are defined.

However, it is often inconvenient to adhere strictly to this convention: in cases where the meaning is fairly obvious from the context the expressions f and $f(t)$ may be used more loosely. For example it is common practice to write t^2 to denote both the number obtained by squaring a given number t and also the function which this operation effectively defines.

1.1.2 Sometimes (though not always) a function is defined by a formula valid for every t within the domain of definition. On the other hand it may happen that the formula becomes meaningless for certain values of t, and it is necessary to complete the definition of the function by assigning specific values to $f(t)$ at such points. Usually we are interested in cases where f is defined for all real numbers t, or else when f is defined for all t in a certain finite interval. In the first place we write

$$y = f(t) \qquad \text{for } -\infty < t < +\infty.$$

In the second it is necessary to distinguish between **open** and **closed** intervals:

   $y = f(t)$ for $a < t < b$, (f is defined on the open interval (a,b)),

   $y = f(t)$ for $a \leqslant t \leqslant b$, ($f$ is defined on the closed interval $[a,b]$).

Examples

(1) $f(t) = t/|t|$. This formula assigns a well-defined real number to each given non-zero value of t; when $t = 0$, however, the formula reduces to the expression $0/0$ which has no meaning. To obtain a function whose domain of definition is the entire real axis we would need to specify the value $f(0)$. The most "natural" way to complete the definition of f would be to write

$$f(t) = t/|t| \quad \text{for all } t \neq 0; \qquad f(0) = 0.$$

Note, however, that other definitions are perfectly possible; if necessary we could always define another function, say f_1, by writing

$$f_1(t) = t/|t| \quad \text{for all } t \neq 0; \qquad f_1(0) = 1.$$

(2) $f(t) = +\sqrt{1 - t^2}$. Here the domain of definition is the closed interval $[-1, +1]$; for all values of t outside this interval, the formula fails to specify any (real) value.

1.2 CONTINUITY

1.2.1 A function f is said to be **continuous** at a point t_0 within its domain of definition if, for all points t sufficiently close to t_0, the functional value $f(t)$ differs from $f(t_0)$ by an arbitrarily small amount. More precisely, let any small, positive number ϵ be given; then we can always find a corresponding positive number η such that whenever $|t - t_0| < \eta$ we must have $|f(t) - f(t_0)| < \epsilon$.

Equivalently we could say that f is continuous at t_0 if and only if

  (i) $f(t_0)$ exists (i.e. t_0 is within the domain of definition of f),
  (ii) $f(t)$ tends to a definite limit as t tends to t_0, and
  (iii) $\lim_{t \to t_0} f(t) = f(t_0)$.

A function f is said to be **continuous in the open interval** (a,b) if it is continuous at each point t of (a,b). If, in addition, $f(t)$ approaches the value $f(a)$ as its limit as t approaches a, and approaches the value $f(b)$ as t approaches b, then f is said to be **continuous on the closed interval** $[a,b]$.

1.2.2 For functions in general (whether continuous or not) the following definitions apply:

  (i) If, for some finite number M, we have $f(t) \leqslant M$ for all t in a certain range of values then f is said to be **bounded above** over that range and M is called an **upper bound** of f. The **smallest** number M for which this inequality holds is called the **least upper bound** (l.u.b.), or the **supremum** (sup.), of f over the range concerned.
  We write

$$M = \text{l.u.b.}\{f(t)\}, \text{ or } M = \sup\{f(t)\} .$$

(ii) If, for some finite number m, we have $f(t) \geqslant m$ for all t in a certain range then f is said to be **bounded below** over that range and m is called a **lower bound** of f. The **largest** number m such that this inequality holds is called the **greatest lower bound** (g.l.b.), or the **infimum** (inf.), of f over the range concerned. We write

$$m = \text{g.l.b.} \{f(t)\}, \text{ or } m = \inf \{f(t)\}.$$

A function f which is continuous on a finite, closed, interval $[a,b]$ actually attains its bounds on that interval, i.e. if f is continuous on $[a,b]$ and if M and m denote respectively its least upper bound and greatest lower bound on that interval, then there exist points t_1 and t_2 in $[a,b]$ such that

$$f(t_1) = M \quad \text{and} \quad f(t_2) = m.$$

This means that for continuous functions we can replace the terms "least upper bound" and "greatest lower bound" by "greatest value of $f(t)$" and "least value of $f(t)$" respectively. Sometimes it is convenient to emphasise this point by using the notations

$$\max_{a \leqslant t \leqslant b} \{f(t)\} \quad \text{instead of} \quad \text{l.u.b} \{f(t)\}_{a \leqslant t \leqslant b}$$

and

$$\min_{a \leqslant t \leqslant b} \{f(t)\} \quad \text{instead of} \quad \text{g.l.b.} \{f(t)\}_{a \leqslant t \leqslant b}$$

Note that if f fails to be continuous at any point in $[a,b]$, or if we drop either the requirement that $[a,b]$ be a finite interval, or that it be closed, then we can no longer assert the existence of a maximum or a minimum value of $f(t)$ in that interval.

1.3 DISCONTINUITIES

1.3.1 Most of the standard functions encountered in elementary calculus are defined by a single formula in each case, valid for all values of t, and are continuous everywhere. In particular this is so for the following functions:

(a) all polynomials; for example, $f(t) = at^2 + bt + c$,
(b) the exponential function e^t, and the hyperbolic functions sinh t and cosh t,
(c) the trigonometric functions sin t and cos t.

A point at which a function fails to be continuous is said to be a **discontinuity** of the function. In elementary calculus such discontinuities as are encountered are usually of a straightforward nature. For example consider the function $f(t) = 1/t$ which is defined and continuous everywhere except at the point $t = 0$. As t approaches 0 from the right-hand side, through positive values, the function becomes indefinitely large and positive; as t approaches 0 from the left, the function becomes indefinitely large and negative. Behaviour of this kind, in which a function becomes unbounded in absolute value in the neighbourhood of a specific point, is typical of discontinuities associated with *rational* functions: a rational function f is a function which can always be expressed as the ratio of two polynomials

$$f(t) = \frac{P(t)}{Q(t)} = \frac{a_n t^n + a_{n-1} t^{n-1} + \ldots + a_0}{b_m t^m + b_{m-1} t^{m-1} + \ldots + b_0}$$

and discontinuities of f will exist precisely at those points $t = t_k$ for which the denominator $Q(t)$ vanishes.

1.3.2 In many applications, however, it is useful to consider functions which remain bounded but which admit sudden, discontinuous, jumps in value. It is precisely when we try to extend the operations and processes of the calculus so as to apply to functions of this type that the need for some generalisation of the concept of function first becomes apparent. This point will be taken up and dealt with in some detail in Chapter 2.

1.4 DIFFERENTIABILITY

1.4.1 Let the function f be defined on the open interval (a,b). The **first derivative** of f is the function f' defined by

$$\lim_{h \to 0} \left[\frac{f(t+h) - f(t)}{h} \right] \equiv f'(t) \equiv \frac{dy}{dt},$$

at all points t of (a,b) for which this limit exists. Note that h is not restricted to positive values, so that the limit is required to be uniquely defined no matter how the point $(t+h)$ approaches t: when this is the case the function f is said to be **differentiable** at the point t. Any function differentiable at a point t must certainly be continuous there. On the other hand f may be continuous at t and yet have no well-defined derivative there.

1.4.2 f is said to be **differentiable on the open interval** (a,b) if $f'(t)$ exists for every t in (a,b). It is said to be **differentiable on the closed interval** $[a,b]$ if, in addition, the following limits exist

$$\lim_{h \to 0} \left[\frac{f(a+h) - f(a)}{h} \right] \quad \text{and} \quad \lim_{h \to 0} \left[\frac{f(b) - f(b-h)}{h} \right]$$

where h goes to 0 through positive values in each case.

If f is differentiable on (a,b) and if f' is continuous on this interval, then f is said to be **continuously differentiable**, or **smooth**, on (a,b). If both f and its first derivative f' are continuous on an interval, save possibly for a finite number of jump discontinuities, then f is said to be **sectionally smooth** on that interval.

A function f whose first derivative f' is itself differentiable on (a,b) is said to be **twice differentiable** on that interval; the first derivative of the function f' is said to be the **second derivative** of the function f, and is written as f''. If $y = f(t)$ then it is usual to write

$$f''(t) = \frac{d^2 y}{dt^2} .$$

Derivatives of higher order are defined similarly.

1.4.3 For differentiable functions in general the following results hold:

(i) If u and v are differentiable functions, and a and b are constants, then $w = au + bv$ is differentiable and

$$\frac{d}{dt}(au + bv) = a\left(\frac{du}{dt}\right) + b\left(\frac{dv}{dt}\right).$$

(ii) If u and v are differentiable then so also is the product function uv and

$$\frac{d}{dt}(uv) = v\frac{du}{dt} + u\frac{dv}{dt} .$$

(iii) If $y = f(x)$ where $x = g(t)$ then, provided the derivatives concerned do exist, we have

$$\frac{dy}{dt} = \frac{dy}{dx}\frac{dx}{dt} .$$

(iv) If $y = f(t)$ is a differentiable function with a well-defined inverse function, $t = f^{-1}(y)$, then

$$\frac{dy}{dt} = 1 \Big/ \frac{dt}{dy} .$$

(v) *Rolle's Theorem.* Let f be continuous on the closed interval $[a,b]$ and differentiable on the open interval (a,b). If $f(a) = f(b) = 0$, then there exists a point t_0 in (a,b) such that $f'(t_0) = 0$.

1.5 TAYLOR'S THEOREM

1.5.1 The so-called **mean value theorems** of the differential calculus are more or less direct consequences of Rolle's Theorem. In view of the extreme importance of these results, and of the consequences which can be deduced from them, we give brief indications of how they may be established:

1.5.2 First Mean Value Theorem

If f is a function which is continuous on the closed interval $[a,b]$ and differentiable on the open interval (a,b) then there exists a point t_0 in (a,b) such that

$$f(b) - f(a) = (b - a)f'(t_0) .$$

Proof. The proof is almost obvious from Fig. 1.1.

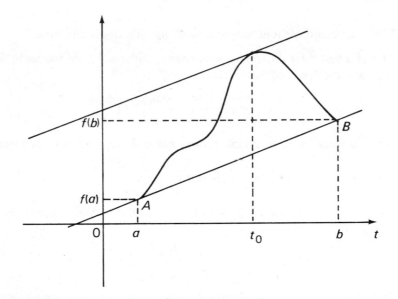

Fig. 1.1.

The equation of the chord AB is given by

$$y = \left[\frac{f(b) - f(a)}{b - a}\right] t + \frac{bf(a) - af(b)}{b - a}$$

or,

$$y = \left[\frac{f(b) - f(a)}{b - a}\right](t - a) + f(a) .$$

Let $F(t) \equiv f(t) - y$; then $F(b) = F(a) = 0$, and the result follows immediately on applying Rolle's Theorem to the function F. ∎

1.5.3 Mean Value Theorem of Order n (Taylor's Theorem)

If the function f, together with its first $(n-1)$ derivatives $f', f'', \ldots, f^{(n-1)}$, are each continuous on the closed interval $[a,b]$, and if the nth derivative $f^{(n)}$ exists on the open interval (a,b), then there exists a point t_n in (a,b) such that

$$f(b) - f(a) = (b-a)f'(a) + \frac{(b-a)^2}{2!}f''(a) + \ldots + \frac{(b-a)^{n-1}}{(n-1)!}f^{(n-1)}(a)$$

$$+ \frac{(b-a)^n}{n!}f^{(n)}(t_n) \,.$$

Alternatively if we put $b = a + h$ then the result can be written in the form,

$$f(a+h) = f(a) + hf'(a) + \frac{h^2}{2!}f''(a) + \ldots + \frac{h^{n-1}}{(n-1)!}f^{(n-1)}(a) + R_n$$

where $R_n = \dfrac{h^n}{n!}f^{(n)}(a + \theta_n h)$, and θ_n is some number lying between 0 and 1.

Proof. The proof is a straightforward generalisation of the first mean value theorem given above, although in the general case there is no simple and obvious geometric interpretation. We have only to apply Rolle's Theorem to the function

$$F_n(t) - \left(\frac{b-t}{b-a}\right)^n F_n(a)$$

where $F_n(t) = f(b) - f(t) - (b-t)f'(t) - \frac{(b-t)^2}{2!}f''(t) - \ldots$

$$- \frac{(b-t)^{n-1}}{(n-1)!}f^{(n-1)}(t) \,.$$

∎

1.5.4 Taylor's Series

Suppose now that the conditions for Taylor's Theorem apply to a function f over an interval (which may be finite or infinite). Let a be a fixed point in that interval and write $t = a + h$ for any other point t within that interval. Then Taylor's Theorem states that

$$f(t) = f(a) + (t-a)f'(a) + \frac{(t-a)^2}{2!}f''(a) + \ldots$$

$$+ \frac{(t-a)^{n-1}}{(n-1)!}f^{(n-1)}(a) + R_n(t) \,.$$

If, in addition, the following two conditions are fulfilled,

   (i) f can be differentiated as often as we wish over the interval concerned
       (f is **infinitely differentiable** there),
   (ii) for each t in the interval concerned we have

$$\lim_{n \to \infty} R_n(t) = 0,$$

then we can allow n to tend to infinity in Taylor's Theorem and obtain the so-called **Taylor expansion of f about the point $t = a$:**

$$f(t) = f(a) + (t-a)f'(a) + \dots + \frac{(t-a)^n}{n!}f^{(n)}(a) + \dots$$

In the particular case $a = 0$ we get the **Maclaurin series** for f:

$$f(t) = f(0) + tf'(0) + \frac{t^2}{2!}f''(0) + \dots + \frac{t^n}{n!}f^{(n)}(0) + \dots$$

1.5.5   The reader is reminded that, among the standard elementary functions, the following expansions are particularly important:

$$e^t = 1 + t + \frac{t^2}{2!} + \frac{t^3}{3!} + \frac{t^4}{4!} + \dots$$

$$\sinh t = t + \frac{t^3}{3!} + \frac{t^5}{5!} + \frac{t^7}{7!} + \dots$$

$$\cosh t = 1 + \frac{t^2}{2!} + \frac{t^4}{4!} + \frac{t^6}{6!} + \dots$$

$$\sin t = t - \frac{t^3}{3!} + \frac{t^5}{5!} - \frac{t^7}{7!} + \dots$$

$$\cos t = 1 - \frac{t^2}{2!} + \frac{t^4}{4!} - \frac{t^6}{6!} + \dots$$

(Each of the above expansions is valid for all values of t.)

$$(1 + t)^\alpha = 1 + \alpha t + \frac{\alpha(\alpha - 1)}{2!}t^2 + \frac{\alpha(\alpha - 1)(\alpha - 2)}{3!}t^3 + \dots$$

(If α is a positive integer the series terminates and reduces to the ordinary binomial expansion. In every other case the expansion is an infinite series valid for all t such that $-1 < t < +1$.)

$$\log(1 + t) = t - \frac{t^2}{2} + \frac{t^3}{3} - \frac{t^4}{4} + \dots \ , \qquad -1 < t \leqslant +1,$$

$$\log(1 - t) = - t - \frac{t^2}{2} - \frac{t^3}{3} - \frac{t^4}{4} - \dots \ , \qquad -1 \leqslant t < +1,$$

$$\log\left[\frac{1 + t}{1 - t}\right] = 2\{t + \frac{t^3}{3} + \frac{t^5}{5} + \frac{t^7}{7} + \dots\} \ , \qquad -1 < t < +1,$$

$$\tan^{-1} t \ \ = t - \frac{t^3}{3} + \frac{t^5}{5} - \frac{t^7}{7} + \dots \ , \qquad -1 \leqslant t \leqslant +1.$$

1.5.6 l'Hopital's Rule

To evaluate limits of the form

$$\lim_{t \to t_0} [f(t)/g(t)] \ \text{ where we have } \ \lim_{t \to t_0} f(t) = \lim_{t \to t_0} g(t) = 0$$

we can often make use of a simple corollary of Taylor's Theorem known as **l'Hopital's Rule**. Suppose that f and g have continuous derivatives in some neighbourhood of the point t_0 and that $f(t_0) = g(t_0) = 0$ while $g'(t_0) \neq 0$. By Taylor's Theorem (with $n = 1$)

$$\frac{f(t_0 + h)}{g(t_0 + h)} = \frac{f(t_0) + hf'(t_0 + \theta h)}{g(t_0) + hg'(t_0 + \theta'h)} = \frac{f'(t_0 + \theta h)}{g'(t_0 + \theta'h)}$$

and so

$$\lim_{t \to t_0} \frac{f(t)}{g(t)} = \lim_{h \to 0} \frac{f(t_0 + h)}{g(t_0 + h)} = \lim_{h \to 0} \frac{f'(t_0 + \theta h)}{g'(t_0 + \theta'h)} = \frac{f'(t_0)}{g'(t_0)}$$

In fact a stronger and more useful form of the rule can be established:

$$\text{if } f(t) \to 0 \text{ and } g(t) \to 0 \text{ as } t \to t_0 \text{ then}$$

$$\lim_{t \to t_0} \frac{f(t)}{g(t)} = \lim_{t \to t_0} \frac{f'(t)}{g'(t)}$$

provided the latter limit exists.

In this form it is not necessary to assume that $f(t_0)$ and $g(t_0)$ are actually defined.

1.6 INTEGRATION

1.6.1 Let f be a function of the real variable t which is continuous for all t in the closed interval $[a,b]$. Subdivide this interval into n sub-intervals by arbitrarily chosen points $t_1, t_2, \ldots, t_n$, and write $t_0 = a, t_n = b$. In each sub-interval $[t_{k-1}, t_k]$ choose some point ξ_k arbitrarily and form the sum

$$f(\xi_1)(t_1 - a) + f(\xi_2)(t_2 - t_1) + \ldots + f(\xi_n)(b - t_{n-1}) \equiv \sum_{k=1}^{n} f(\xi_k)\Delta_k .$$

Now let the number of subdivisions n increase in such a way that, for each k, the number $\Delta_k \equiv t_k - t_{k-1}$ tends to zero. Then it can be shown that the sums tend to a unique limit (independently of how the points of subdivision t_k, and the points ξ_k of the resulting sub-intervals, have been chosen). This limit is, by definition, the elementary, or **Riemann**, integral of the continuous function f from $t = a$ to $t = b$. More precisely given any number $\epsilon > 0$, however small, we can always find a positive number h such that

$$\left| \int_a^b f(t)\,dt - \sum_{k=1}^{n} f(\xi_k)\Delta_k \right| < \epsilon$$

for any sum of the form described above in which $\Delta_k < h$ for every k.

1.6.2 As is well known, the geometrical significance of the integral lies in the fact that it gives a measure of the area enclosed by the graph of the function $y = f(t)$, the ordinates $t = a$ and $t = b$, and the t-axis (with the convention that areas lying above the axis are taken to be positive while those lying below are taken to be negative). The other important aspect of integration is its role as, in some sense, a process inverse to differentiation. This connection follows as a result of the **first mean value** theorem of the integral calculus:

First Mean Value Theorem (for integrals). If f is continuous on the closed interval $[a,b]$ then there must exist some point ξ in (a,b) such that

$$(b - a)f(\xi) = \int_a^b f(t)\,dt.$$

Proof. As in the case of the first mean value theorem of the differential calculus, the geometrical significance of the theorem (Fig. 1.2) is almost obvious.

Let m, M denote respectively the least and the greatest of the values which the continuous function f assumes in the interval $[a,b]$. Then for any sub-division by points t_k we have

$$m\Delta_k \leqslant f(\xi_k)(t_k - t_{k-1}) \leqslant M\Delta_k .$$

Hence,

$$m \sum_{k=1}^{n} (t_k - t_{k-1}) \leqslant \sum_{k=1}^{n} f(\xi_k)(t_k - t_{k-1}) \leqslant M \sum_{k=1}^{n} (t_k - t_{k-1})$$

so that, in the limit,

$$m(b-a) \leqslant \int_a^b f(t)\,dt \leqslant M(b-a) \,.$$

Since f is continuous on $[a,b]$ it takes on every value between m and M, its least and greatest values there. Hence there must exist some point ξ in (a,b) such that

$$\frac{1}{b-a}\int_a^b f(t)\,dt = f(\xi) \,.$$

■

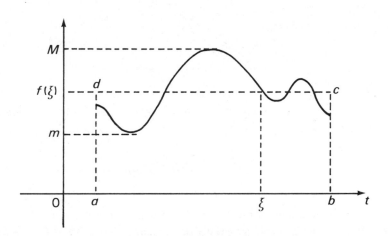

Fig. 1.2 — Area of rectangle abcd = $(b-a)f(\xi)$.

1.6.3 Fundamental Theorem of the Calculus

If f is continuous on $[a,b]$, define a function F on $[a,b]$ by writing

$$F(t) = \int_a^t f(\tau)\,d\tau \,.$$

Then,

$$\frac{F(t+h)-F(t)}{h} = \frac{1}{h}\left[\int_a^{t+h} f(\tau)\,d\tau - \int_a^t f(\tau)\,d\tau\right]$$

$$= \frac{1}{h}\int_t^{t+h} f(\tau)\,d\tau = f(\xi), \text{ for some } \xi \text{ in } (t,t+h) \,.$$

If we allow h to tend to 0 then, since f is continuous, $f(\xi)$ tends to $f(t)$ as its limit, and so

$$F'(t) = \lim_{h \to 0} \frac{F(t+h) - F(t)}{h} = f(t) \ .$$

Now let F denote *any* function whose derivative is f. Then we can always write

$$F(t) = \int_a^t f(\tau)\,d\tau + C$$

where C is a constant.

Putting $t = a$ in this gives

$$F(a) = \int_a^a f(\tau)\,d\tau + C = C.$$

Hence, putting $t = b$, we obtain the result usually described as the Fundamental Theorem of the Calculus:

$$\int_a^b f(\tau)\,d\tau = F(b) - F(a)$$

where F is any function whose derivative is equal to the integrand f on $[a,b]$. (Such a function is often called a **primitive**, or an **anti-derivative** of f on $[a,b]$.)

1.6.4 Remark

The definition of the integral as the limit of finite sums is not confined to continuous functions. If f is defined and bounded on $[a,b]$ then we can always construct approximating sums of the form

$$\sum_{k=1}^n f(\xi_k)\Delta_k,$$

but without the hypothesis of continuity it does not necessarily follow that these sums converge to a definite limit as the sub-divisions are taken smaller and smaller. Whenever such a limit does exist the function f, whether continuous or not, is said to be **integrable in the Riemann sense**, [integrable-R] over $[a,b]$. This will be the case, for example, when f has nothing worse than finitely many jump discontinuities on $[a,b]$. Note that whenever f is Riemann-integrable over $[a,b]$ we can always define the function

$$F(t) = \int_a^t f(\tau)\mathrm{d}\tau .$$

Further,

$$|F(t+h) - F(t)| = \left| \int_t^{t+h} f(\tau)\mathrm{d}\tau \right| < |h| . \sup_{t \leqslant \tau \leqslant t+h} |f(\tau)| .$$

Since f is bounded on $[a,b]$ it follows that $|F(t+h) - F(t)|$ tends to zero with h, i.e. the *function F is continuous on $[a,b]$ whether the integrand f is itself continuous or not.*

1.7 IMPROPER INTEGRALS

1.7.1 Suppose that f is bounded and continuous (more generally, bounded and Riemann-integrable) over every interval $[a,t]$, where a is some fixed number and t is any number greater than a. The **infinite**, or **improper**, Riemann integral $\int_a^\infty f(\tau)\mathrm{d}\tau$ is defined by the relation

$$\int_a^\infty f(\tau)\mathrm{d}\tau \equiv \lim_{t \to \infty} \int_a^t f(\tau)\mathrm{d}\tau = \lim_{t \to \infty} F(t) - F(a)$$

where F denotes any primitive of f. If the limit is finite, the integral is said to **converge**; otherwise it is said to **diverge**. An integral of this type, in which the range of integration becomes infinite but the integrand remains bounded, is called an **improper integral of the first kind**. More generally we write

$$\int_{-\infty}^{+\infty} f(\tau)\mathrm{d}\tau \equiv \lim_{t_1 \to \infty} \int_{-t_1}^{a} f(\tau)\mathrm{d}\tau + \lim_{t_2 \to \infty} \int_a^{t_2} f(\tau)\mathrm{d}\tau .$$

If $\int_a^\infty |f(\tau)|\,\mathrm{d}\tau$ converges then $\int_a^\infty f(\tau)\,\mathrm{d}\tau$ must also converge; the convergence of $\int_a^\infty f(\tau)\,\mathrm{d}\tau$ in this case is said to be **absolute**. If $\int_a^\infty |f(\tau)|\,\mathrm{d}\tau$ diverges then the integral $\int_a^\infty f(\tau)\,\mathrm{d}\tau$ may diverge also, or it may converge; in the latter case the convergence is said to be **conditional**.

Comparison Test (integrals of the first kind). Let f and g be bounded integrable functions of t for $a \leqslant t \leqslant x$, and be such that $0 \leqslant f(t) \leqslant g(t)$ throughout this range. Then

(i) if $\int_a^\infty g(t)\,\mathrm{d}t$ converges, so also does $\int_a^\infty f(t)\,\mathrm{d}t .$

(ii) if $\int_a^\infty f(t)\,dt$ diverges, so also does $\int_a^\infty g(t)\,dt$.

A particularly useful application of this general test is obtained by taking $g(t) = t^{-p}$. For, if $a > 0$ and $p \neq 1$ then we get

$$\int_a^\infty \frac{dt}{t^p} = \lim_{T \to \infty} \int_a^T \frac{dt}{t^p} = \lim_{T \to \infty} \left[\frac{t^{1-p}}{1-p} \right]_a^T .$$

This converges to the value $a^{1-p}/(p-1)$ if $p > 1$ and diverges to $+\infty$ if $p < 1$.

In case $p = 1$ we have

$$\int_a^\infty \frac{dt}{t} = \lim_{T \to \infty} \int_a^T \frac{dt}{t} = \lim_{T \to \infty} \left\{ \log T - \log a \right\}$$

and this diverges to $+\infty$.

Accordingly we derive the simple and important "p-test" for improper integrals of the first kind:

Let f be a bounded continuous function which is non-negative for all $t \geqslant a$ (where $a > 0$). If there exists a number p such that $\lim_{t \to a} t^p f(t) = A$ then,

(i) $\int_a^\infty f(t)\,dt$ converges if $p > 1$ and A is finite.

(ii) $\int_a^\infty f(t)\,dt$ diverges if $p \leqslant 1$ and $A > 0$ (possibly infinite).

1.7.2 Now suppose that f is a function which becomes unbounded as t approaches a in the interval $[a,b]$. We define

$$\int_a^b f(t)\,dt \equiv \lim_{\epsilon \downarrow 0} \int_{a+\epsilon}^b f(t)\,dt$$

whenever this limit exists. Similarly if f becomes unbounded as t tends to b in $[a,b]$ then we define

$$\int_a^b f(t)\,dt \equiv \lim_{\epsilon \downarrow 0} \int_a^{b-\epsilon} f(t)\,dt .$$

Finally if t_0 is some point in (a,b) and if f becomes unbounded in the neighbourhood of t_0 then the improper integral $\int_a^b f(t)\,dt$ is said to converge to the value

$$\lim_{\epsilon_1 \downarrow 0} \int_a^{t_0 - \epsilon_1} f(t)\,dt + \lim_{\epsilon_2 \downarrow 0} \int_{t_0 + \epsilon_2}^b f(t)\,dt$$

provided that the limits exist *independently* (that is, ϵ_1 and ϵ_2 must be allowed to tend to zero independently of one another; we require that *both* the integrals

$$\int_a^{t_0} f(t)\,dt \quad \text{and} \quad \int_{t_0}^b f(t)\,dt$$

should converge.)

Integrals such as these, in which the range of integration is finite but the integrand becomes unbounded at one or more points of that range, are called **improper integrals of the second kind**. In case we have both an infinite range of integration and an integrand which becomes unbounded within that range, we speak of an **improper (Riemann) integral of the third kind**.

Comparison Test (integrals of the second kind). Let f and g be continuous for $a \leqslant t \leqslant b$ and such that $0 \leqslant f(t) \leqslant g(t)$ throughout this range. Suppose also that both f and g become unbounded as t tends to a. Then

(i) if $\displaystyle\int_a^b g(t)\,dt$ converges, so also does $\displaystyle\int_a^b f(t)\,dt$.

(ii) if $\displaystyle\int_a^b f(t)\,dt$ diverges, so also does $\displaystyle\int_a^b g(t)\,dt$.

Once again we obtain a convergence test of particular importance by taking $g(t) = t^{-p}$. If $p \neq 1$ then

$$\int_0^1 \frac{dt}{t^p} = \lim_{\epsilon \downarrow 0} \int_\epsilon^1 \frac{dt}{t^p} = \frac{1}{1-p} - \lim_{\epsilon \downarrow 0} \left[\frac{\epsilon^{1-p}}{1-p}\right]$$

which converges to the value $1/(1-p)$ if $p < 1$ and diverges to $+\infty$ if $p > 1$. Also if $p = 1$ we have

$$\int_0^1 \frac{dt}{t} = \lim_{\epsilon \downarrow 0} \int_\epsilon^1 \frac{dt}{t} = \log 1 - \lim_{\epsilon \downarrow 0} \left[\log \epsilon\right]$$

and this diverges to $+\infty$.

As a result we obtain a corresponding "p-test" for improper integrals of the second kind:

Let f be a continuous, non-negative, function for $a < t \leqslant b$ and suppose that f becomes unbounded as t tends to a. If there exists a number p such that $\lim_{t \to a} (t-a)^p f(t) = A$ then

(i) $\displaystyle\int_a^b f(t)\,dt$ converges if $p < 1$ and A is finite.

(ii) $\displaystyle\int_a^b f(t)\,dt$ diverges if $p \geqslant 1$ and $A > 0$ (possibly infinite).

1.7.3 Cauchy Principal Value

Suppose that f is a function which becomes unbounded in the neighbourhood of a point t_0 in (a,b) and that the improper integral $\displaystyle\int_a^b f(t)\,dt$ diverges. It may happen that although the limits

$$\lim_{\epsilon \downarrow 0} \int_a^{t_0 - \epsilon_1} f(t)\,dt \quad \text{and} \quad \lim_{\epsilon_2 \downarrow 0} \int_{t_0 + \epsilon_2}^b f(t)\,dt$$

do not exist independently we may set $\epsilon_1 = \epsilon_2 = \epsilon$ and then find that the unbounded parts of the two integrals cancel each other out. This gives a certain finite answer, called the **Cauchy principal value** of the (divergent) integral. We usually write

$$P\int_a^b f(t)\,dt \equiv \lim_{\epsilon \downarrow 0} \left[\int_a^{t_0 - \epsilon} f(t)\,dt + \int_{t_0 + \epsilon}^b f(t)\,dt \right]$$

For example, $P\displaystyle\int_{-1}^{+1} \frac{dt}{t} = 0$, although both the integrals $\displaystyle\int_{-1}^{-\epsilon} \frac{dt}{t}$ and $\displaystyle\int_{\epsilon}^{1} \frac{dt}{t}$

become arbitrarily large in absolute value as ϵ tends to 0.

1.8 NOTE ON UNIFORM CONVERGENCE

1.8.1 A sequence (f_m) of functions converges to a limit function f in the simple, or **pointwise**, sense if it is the case that

$$\lim_{m \to \infty} f_m(t) = f(t)$$

for each point in the domain of definition concerned. This means that given any number $\epsilon > 0$ we can find, for each particular value of t, a corresponding integer $m_0 = m_0(\epsilon, t)$ such that

$$|f(t) - f_m(t)| < \epsilon$$

for all $m \geqslant m_0$. As our notation indicates, the number m_0 will generally depend on the particular value of t concerned as well as on the number ϵ.

Now suppose that (f_m) is a sequence of (bounded) functions which converges to a limit function f in the sense that

$$\lim_{m \to \infty} \left[\sup_t |f(t) - f_m(t)| \right] = 0.$$

This is to say that given any $\epsilon > 0$ we can always find a corresponding integer $m_0 = m_0(\epsilon)$ such that for all $m \geqslant m_0$ we have

$$\sup_t |f(t) - f_m(t)| < \epsilon.$$

It follows that

$$|f(t) - f_m(t)| < \epsilon$$

for all $m \geqslant m_0$ and for *every* t. This time the integer m_0 depends only on ϵ and is independent of t; the convergence of the f_m to f is then said to be **uniform**.

1.8.2 Theorem

If (f_m) is a sequence of bounded, continuous, functions which converges uniformly to f, then f is itself bounded and continuous.

Proof. For any t and x we have

$$|f(t+x) - f(t)| \leqslant |f(t+x) - f_m(t+x)|$$

$$+ |f_m(t+x) - f_m(t)| + |f_m(t) - f(t)|.$$

Given $\epsilon > 0$ we can find a corresponding integer $m_0 = m_0(\epsilon)$ such that for all t and x

$$|f(t+x) - f_{m_0}(t+x)| < \frac{\epsilon}{3} \quad \text{and} \quad |f_{m_0}(t) - f(t)| < \frac{\epsilon}{3}$$

Again, by the continuity of f_{m_0} at any given point t, we can always find $\eta > 0$ such that

$$|f_{m_0}(t+x) - f_{m_0}(t)| < \frac{\epsilon}{3} \quad \text{for all } x \text{ such that } |x| < \eta.$$

It follows at once that f is certainly a continuous function. That it is also bounded is a consequence of the fact that for all t

$$f_{m_0}(t) - \epsilon < f(t) < f_{m_0}(t) + \epsilon.$$ ∎

1.8.3 Theorem

If f is the uniform limit of a sequnce (f_m) of bounded, continuous functions on the finite, closed interval $[a,b]$ then

$$\int_a^b f(t)\,dt = \lim_{m \to \infty} \int_a^b f_m(t)\,dt .$$

Proof. If the sequence (f_m) converges uniformly to f then by the Theorem of 1.8.2, the limit function f is certainly bounded and continuous, and is therefore integrable over $[a,b]$. Also

$$\left| \int_a^b f(t)\,dt - \int_a^b f_n(t)\,dt \right| \leqslant \int_a^b \left| f(t) - f_n(t) \right| dt$$

$$\leqslant (b-a). \sup_t \left| f(t) - f_n(t) \right|$$

and the result follows since $\displaystyle\lim_{n \to \infty} \left[\sup \left| f(t) - f_n(t) \right| \right] = 0 .$

∎

Corollary

Suppose further that each of the derivatives f_m' exists and is continuous on $[a,b]$ and that the sequence (f_m') converges uniformly to a function ϕ on $[a,b]$. Then $\phi(t) = f'(t)$.

Proof. The uniformity of the convergence shows that ϕ must be continuous on $[a,b]$. For any t in $[a,b]$ we have

$$\int_a^t \phi(\tau)\,d\tau = \lim_{n \to \infty} \int_a^t f_n'(\tau)\,d\tau = \lim_{n \to \infty} \left[f_n(t) - f_n(a) \right] = f(t) - f(a) .$$

The Fundamental Theorem of the Calculus then shows that f is differentiable on $[a,b]$ and that $f'(t) = \phi(t)$.

∎

1.9 NOTE ON LEIBNITZ'S RULE FOR DIFFERENTIATING AN INTEGRAL

1.9.1 From the Fundamental Theorem of the Calculus we know that if f is a function bounded and continuous on the finite closed interval $[a,b]$ then

$$\int_a^b f(t)\,dt = F(b) - F(a)$$

where F is a function such that $F'(t) = f(t)$ on $[a,b]$. Hence if either the top limit b or the bottom limit a is allowed to vary then it follows that

$$\frac{\mathrm{d}}{\mathrm{d}b}\left[\int_a^b f(t)\,\mathrm{d}t\right] = \frac{\mathrm{d}}{\mathrm{d}b}\left[F(b) - F(a)\right] = f(b)$$

and

$$\frac{\mathrm{d}}{\mathrm{d}a}\left[\int_a^b f(t)\,\mathrm{d}t\right] = \frac{\mathrm{d}}{\mathrm{d}a}\left[F(b) - F(a)\right] = -f(a) .$$

1.9.2 Now let $f(x,y)$ be a function defined and continuous throughout some region R of the (x,y)-plane and such that both the partial derivatives $\partial f/\partial x$ and $\partial f/\partial y$ exist and are continuous at each point of R. Let the functions $a = \phi_1(x)$ and $b = \phi_2(x)$ have continuous derivatives throughout the interval $a \leqslant x \leqslant b$, and consider the function

$$G(x) \equiv \int_a^b f(x,y)\,\mathrm{d}y \equiv G(a,b,x) .$$

By the chain rule for partial derivatives we have

$$\frac{\mathrm{d}G}{\mathrm{d}x} = \frac{\partial G}{\partial x} + \frac{\partial G}{\partial a}\frac{\mathrm{d}a}{\mathrm{d}x} + \frac{\partial G}{\partial b}\frac{\mathrm{d}b}{\mathrm{d}x}$$

Then

(i) $$\frac{\partial G}{\partial a} = \frac{\partial}{\partial a}\int_a^b f(x,y)\,\mathrm{d}y = -f(x,a)$$

(ii) $$\frac{\partial G}{\partial b} = \frac{\partial}{\partial b}\int_a^b f(x,y)\,\mathrm{d}y = f(x,b)$$

(iii) $$\frac{\partial G}{\partial x} \equiv \lim_{h \to 0}\left[\int_a^b f(x+h,y)\,\mathrm{d}y - \int_a^b f(x,y)\,\mathrm{d}y\right]\Bigg/h$$

$$= \lim_{h \to 0}\int_a^b \frac{1}{h}\left[f(x+h,y) - f(x,y)\right]\mathrm{d}y = \lim_{h \to 0}\int_a^b f_x(x+\theta h,y)\,\mathrm{d}y$$

where $0 < \theta < 1$. Since $f_x(x,y) \equiv \partial f/\partial x$ is assumed to be continuous in $[a,b]$ we can write

$$f_x(x+\theta h,y) = f_x(x,y) + \epsilon$$

where $|\epsilon|$ tends to 0 with h. It follows that

$$\frac{\partial G}{\partial x} = \int_a^b \frac{\partial}{\partial x} f(x,y) \mathrm{d}y \ .$$

Thus finally we obtain what is known as **Leibnitz's Rule** for differentiating an integral with respect to a parameter:

$$\frac{\mathrm{d}}{\mathrm{d}x} \int_{\phi_1(x)}^{\phi_2(x)} f(x,y) \mathrm{d}y = \int_{\phi_1(x)}^{\phi_2(x)} \frac{\partial}{\partial x} f(x,y) \mathrm{d}y$$

$$-\phi_1'(x) f\{x, \phi_1(x)\} + \phi_2'(x) f\{x, \phi_2(x)\} \ .$$

CHAPTER 2

The Delta Function

2.1 FUNCTIONS WITH SIMPLE DISCONTINUITIES

2.1.1 If t_0 is any point within the domain of definition of a function f then any point t *to the right* of t_0 can be expressed as $t = t_0 + \eta$, where $\eta > 0$. As η tends to zero the point $(t_0 + \eta)$ approaches t_0 from the right; if the corresponding values $f(t_0 + \eta)$ tend to a limiting value then this value is called the **right-hand limit** of f as t tends to t_0 and is usually written as $f(t_0+)$:

$$f(t_0+) = \lim_{\eta \to 0} f(t_0 + \eta) \quad , \quad \eta > 0. \tag{2.1}$$

If, in addition, this limit is equal to the value, $f(t_0)$, which the function f actually assumes at the point t_0 then f is said to be **continuous from the right** at t_0.

Similarly any point *to the left* of t_0 can be expressed as $t = t_0 - \eta$ where again η is positive. The limit, if it exists, of $f(t_0 - \eta)$ as η tends to 0 through positive values is called the **left-hand limit** of f as t tends to t_0 and is usually written as $f(t_0-)$. If this limit does exist and is equal to the value $f(t_0)$, then f is said to be **continuous from the left** at t_0.

Clearly f is continuous at t_0 in the usual sense if and only if it is both continuous from the right and continuous from the left there:

$$\lim_{\eta \to 0} f(t_0 + \eta) = \lim_{\eta \to 0} f(t_0 - \eta) = f(t_0). \tag{2.2}$$

If f is continuous at a point t_0 then it must certainly be the case that $f(t_0)$ is defined. A comparatively trivial example of a discontinuity is afforded by a formula like

$$\frac{\sin t}{t}$$

which defines a function continuous everywhere except at the point $t = 0$ where the formula becomes meaningless. We have

$$f(0+) = \lim_{\eta \downarrow 0} \frac{\sin \eta}{\eta} = 1 \quad \text{and} \quad f(0-) = \lim_{\eta \downarrow 0} \frac{\sin (-\eta)}{(-\eta)} = 1$$

and so the discontinuity may be easily removed by *defining* $f(0)$ to be 1. In general a function f is said to have a **removable discontinuity** at t_0 if both the right-hand and the left-hand limits of f at t_0 exist and are equal but $f(t_0)$ is either undefined or else has a value different from $f(t_0+)$ and $f(t_0-)$. The discontinuity disappears on suitably defining (or re-defining) $f(t_0)$. If the one-sided limits $f(t_0+)$ and $f(t_0-)$ both exist but are unequal in value then f is said to have a **simple**, or **jump**, discontinuity at t_0. The number $f(t_0+) - f(t_0-)$ is then called the **saltus**, or the **jump**, of the function at t_0. The simplest example is such a discontinuity occurs in the definition of a special function which is of particular importance.

We shall define **the unit step function**, u, as the function which is equal to 1 for every positive value of t and equal to 0 for every negative value of t. This could as well be defined in terms of a specific formula; for example,

$$u(t) = \frac{1}{2}\left[1 + \frac{t}{|t|} \right] = \begin{cases} 1 \text{ for } t > 0. \\ 0 \text{ for } t < 0. \end{cases} \tag{2.3}$$

The value $u(0)$ is left undefined; in many contexts, as will become clear later, this does not matter. However, some texts do specify a particular value for $u(0)$; the most popular candidates are $u(0) = 0$, $u(0) = 1$, and $u(0) = \frac{1}{2}$, and each value has its defenders. Where necessary we shall adopt the following convention.

The symbol u will always refer to the unit step function, defined as above, with $u(0)$ left unspecified; for any given real number c we shall understand by u_c the function defined by

$$u_c(t) = \begin{cases} 1 & \text{for } t > 0 \\ c & \text{for } t = 0 \\ 0 & \text{for } t < 0. \end{cases} \tag{2.4}$$

Note that taking $c = 0$ gives a function u_0 which is continuous from the left at the origin; similarly, taking $c = 1$ gives a function u_1 which is continuous from the right there.

Figure 2.1 shows the graph of $y = u(t)$, together with those of the functions $u(t - a)$ and $u(a - t)$ obtained by translation and reflection.

2.1.2 The importance of the unit step function can be gauged from the following considerations: suppose that ϕ is a function which is continuous everywhere except for the point $t = a$, at which it has a simple discontinuity (Fig. 2.2).

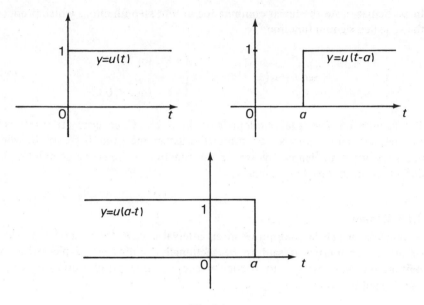

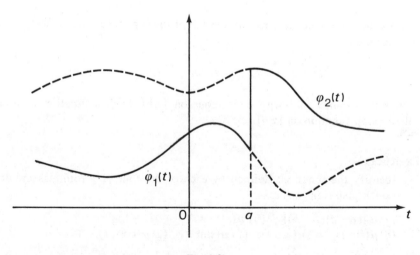

Fig. 2.1.

Fig. 2.2.

Then ϕ can always be represented as a linear combination of functions ϕ_1 and ϕ_2 which are continuous everywhere but which have been truncated at $t = a$:

$$\phi(t) = \phi_1(t)u(a-t) + \phi_2(t)u(t-a) = \begin{cases} \phi_1(t) & \text{for all } t < a \\ \\ \phi_2(t) & \text{for all } t > a . \end{cases} \qquad (2.5)$$

In particular, note the linear combination of unit step functions which produces the so-called **signum function**:

$$\mathrm{sgn}\, t = u(t) - u(-t) = \begin{cases} +1 & \text{for } t > 0 \\ -1 & \text{for } t < 0 \ . \end{cases} \qquad (2.6)$$

The signum function is left undefined at the origin. Once again usage differs in this respect, and a number of authorities assign the value 0 to sgn 0; where necessary we shall denote by $\mathrm{sgn}_0\, t$ the function obtained by completing the definition of sgn t in this manner.

2.1.3 Remark

A function which is continuous in an interval except for a finite number of simple discontinuities is said to be **sectionally continuous** or **piecewise continuous** on that interval. In particular suppose that a finite interval $[a,b]$ is sub-divided by points $t_0, t_1, t_2, \ldots, t_n$ where

$$a = t_0 \leqslant t_1 \leqslant t_2 \leqslant \ldots \leqslant t_n = b \ .$$

If s is a function which is constant on each of the (open) sub-intervals (t_{k-1}, t_k), for example if

$$s(t) = \alpha_k \text{ for } t_{k-1} < t < t_k \qquad (k = 1, 2, \ldots, n),$$

then s is certainly piecewise continuous on $[a,b]$. Such a function is usually called a **step-function** on $[a,b]$.

Exercises I

1. Identify the functions defined by each of the following formulas and draw sketch graphs in each case:

   (a) $u\{(t-a)(t-b)\}$; (b) $u(e^t - \pi)$; (c) $u(t - \log \pi)$;
   (d) $u(\sin t)$; (e) $u(\cos t)$; (f) $u(\sinh t)$; (g) $u(\cosh t)$.

2. Identify the functions defined by each of the following formulas and draw sketch graphs in each case:

   (a) $\mathrm{sgn}\, (t^2 - 1)$; (b) $\mathrm{sgn}\, (e^{-t})$; (c) $\mathrm{sgn}\, (\tan t)$; (d) $\mathrm{sgn}\, (\sin \frac{1}{t})$,
   (e) $t^2 \,\mathrm{sgn}\, t$; (f) $\sin t . \{\mathrm{sgn}\, (\sin t)\}$; (g) $\sin t . \{\mathrm{sgn}\, (\cos t)\}$.

3. A function g is said to be **absolutely continuous** if the following criterion is satisfied; there exists a function h which is integrable over every finite

interval $[a,b]$ within the domain of definition of g and which is such that

$$g(b) - g(a) = \int_a^b h(t)\,dt \ .$$

Give an example to show that an absolutely continuous function need not be differentiable everywhere.

4. A whole family of useful and important functions is generated by replacing the variable t by $|t|$ in many of the familiar formulas encountered in elementary calculus. The functions so obtained are generally differentiable except at isolated points where the derivatives have jump discontinuities. Find the derivatives of each of the functions listed below and sketch the graphs of the function and its derivative in each case. Are these functions absolutely continuous in the sense of question (3) above?

(a) $1 - |t|^3$; (b) $e^{|t|}$; (c) $e^{-|t|}$; (d) $\sin|t|$; (e) $|\sin t|$; (f) $\left|\sin|t|\right|$; (g) $\sinh|t|$.

5.* The examples of the preceding question should make it clear that *continuity* does not imply *differentiability*. Although the reverse implication is in fact valid, it is instructive to realise that the mere existence of a derivative at every point is no guarantee of "reasonable" behaviour:

(a) Prove that if f is differentiable at a point t_0 then it must necessarily be continuous there.
(b) Let $f(t) = t^2.\sin 1/t$ for $t \neq 0$, and $f(0) = 0$. Show that f is differentiable everywhere but that the derivative f' is neither continuous from the right nor continuous from the left at the origin.
(c) Let $g(t) = t^2.\sin 1/t^2$ for $t \neq 0$, and $g(0) = 0$. Show that g is differentiable everywhere, but that its derivative g' is not even bounded in the neighbourhood of the origin.

6. Find each of the following limit functions:

(a) $\lim\limits_{n \to \infty}\left[\dfrac{1}{2} + \dfrac{1}{\pi}\tan^{-1}nt\right]$, where $\tan^{-1} t$ is understood to denote the principal value of the inverse tangent. viz.: $-\pi/2 < \tan^{-1} t \leqslant \pi/2$.
(b) $\lim\limits_{n \to \infty}\{\exp(-e^{-nt})\}$.

2.2 DERIVATIVE OF THE UNIT STEP FUNCTION

2.2.1 Suppose that we try to extend the definition of differentiation in such a way that it applies to functions with jump discontinuities. In particular we would need to define a "derivative" for the unit step function, u. For all $t \neq 0$ this is of course well-defined:

$$u'(t) = 0 \quad \text{for all } t \neq 0 ,$$

corresponding to the obvious fact that the graph of $y = u(t)$ has zero slope for all non-zero values of t. At $t = 0$, however, there is a jump discontinuity, and the definition of derivative accordingly fails. A glance at the graph suggests that it would not be unreasonable to describe the slope as "infinite" at this point; moreover, if we take any specific representation, u_c, of the unit step, then it is certainly the case that

$$\lim_{h \to 0} \frac{u_c(h) - u_c(0)}{h} = +\infty .$$

Thus, from a descriptive point of view at least, the derivative of u would appear to be a function, δ, which has the following pointwise specification:

$$\delta(t) \equiv u'(t) = \begin{cases} 0 & \text{for all } t \neq 0 \\ +\infty & \text{for } t = 0 \end{cases} \tag{2.7}$$

Now let f be any function continuous on a neighbourhood of the origin, say for $-a < t < +a$. Then we should have

$$\int_{-a}^{+a} f(t)u'(t)\mathrm{d}t = \int_{-a}^{+a} f(t) \lim_{h \to 0} \left[\frac{u(t+h) - u(t)}{h} \right] \mathrm{d}t .$$

Assuming that it is permissible to interchange the operations of integration and of taking the limit as h tends to 0, this gives

$$\int_{-a}^{+a} f(t)\delta(t)\mathrm{d}t = \lim_{h \to 0} \frac{1}{h} \int_{-h}^{0} f(t)\mathrm{d}t = \lim_{h \to 0} f(\xi)$$

where ξ is some point lying between $-h$ and 0 (using the First Mean Value Theorem of the Integral Calculus). Since f is assumed continuous in the neighbourhood of $t = 0$, it follows that

$$\int_{-a}^{+a} f(t)\delta(t)\mathrm{d}t = f(0) \tag{2.8}$$

2.2.2 Remark

If f happens to be continuously differentiable in the neighbourhood of the origin then the same result can be obtained by a formal application of integration by parts (although the critical importance of the order in which the operations of integration and of proceeding to the limit are carried out is not made manifest):

$$\int_{-a}^{+a} f(t)\delta(t)\,\mathrm{d}t = \int_{-a}^{+a} f(t)u'(t)\,\mathrm{d}t$$

$$= [u(t)f(t)]_{-a}^{+a} - \int_{-a}^{+a} f'(t)u(t)\,\mathrm{d}t$$

$$= f(a) - \int_{0}^{a} f'(t)\,\mathrm{d}t = f(a) - \{f(a) - f(0)\} = f(0)\ .$$

2.2.3 The result (2.8) is usually referred to as the **sampling property** of the function δ; following the usual convention we shall refer to the function δ as the **delta function**, or the **Dirac delta function**. The sampling property itself is clearly independent of the actual value of a and depends only on the behaviour of the integrand at (or near) the point $t = 0$; accordingly it is most usually stated in the following form:

The sampling property of the delta function: If f is any function which is continuous on a neighbourhood of the origin, then

$$\int_{-\infty}^{+\infty} f(t)\delta(t)\,\mathrm{d}t = f(0) \qquad (2.9)$$

In particular, let $f(t) = u(a - t)$, where a is any fixed number other than 0. Then $f(t) = 0$ for all $t > a$ and $f(t) = 1$ for all $t < a$; moreover f is certainly continuous on a neighbourhood of the origin. Hence we have

$$\int_{-\infty}^{a} \delta(t)\,\mathrm{d}t = \int_{-\infty}^{+\infty} u(a - t)\delta(t)\,\mathrm{d}t = u(a), \quad \text{for any } a \neq 0\ .$$

This shows that the relation $u'(t) = \delta(t)$, from which we first started, could have been deduced as a consequence of the sampling property.

Exercises II

1. Show that δ behaves formally like an *even* function: $\delta(-t) = \delta(t)$.
2. If a is any fixed number and f any function continuous on some neighbourhood of a, show that

$$\int_{-\infty}^{+\infty} f(t)u'(t - a)\,\mathrm{d}t = f(a)\ .$$

3. Obtain the sampling properties which should be associated with each of the following functions:

(a) the derivative of $u(-t)$; (b) the derivative of sgn t.

2.3 THE POINTWISE BEHAVIOUR OF THE DELTA FUNCTION

2.3.1 It is not difficult to show that in point of fact no function can exist which enjoys the properties attributed above to δ. The requirement that $\delta(t) = u'(t)$ necessarily implies that $\delta(t) = 0$ for all $t \neq 0$. Hence, for any continuous function $f(t)$, and any positive numbers ϵ_1, ϵ_2, however small, we must have

$$\int_{-\infty}^{-\epsilon_1} f(t)\delta(t)\mathrm{d}t = \int_{\epsilon_2}^{+\infty} f(t)\delta(t)\mathrm{d}t = 0$$

so that

$$\int_{-\infty}^{+\infty} f(t)\delta(t)\mathrm{d}t = \lim_{\epsilon_1 \downarrow 0} \int_{-\infty}^{-\epsilon_1} f(t)\delta(t)\mathrm{d}t + \lim_{\epsilon_2 \downarrow 0} \int_{\epsilon_2}^{+\infty} f(t)\delta(t)\mathrm{d}t = 0 \ .$$

This contradicts the sampling property (2.9). In particular, for any $t > 0$, we have

$$\int_{-\infty}^{t} \delta(\tau)\mathrm{d}\tau = \int_{-\infty}^{t} u'(\tau)\mathrm{d}\tau = 0 \neq u(t)$$

which is not consistent with the initial assumption that δ is to behave as the derivative of u.

2.3.2 One way of dealing with this situation is to construct a suitable sequence of functions which approximate the desired behaviour of δ, and to treat all expressions involving delta functions in the sense of a symbolic shorthand for certain limiting processes. As a simple example consider sequences $\{s_n\}$ and $\{d_n\}$ defined as follows (see Figs. 2.3(a) and 2.3(b)):

$$s_n(t) = \begin{cases} 1 & , \quad t > 1/2n \\ nt + \dfrac{1}{2}, & -\dfrac{1}{2n} \leqslant t \leqslant \dfrac{1}{2n} \ ; \\ 0 & , \quad t < -1/2n \end{cases} \qquad d_n(t) = \begin{cases} 0, & t > 1/2n \\ n, & -\dfrac{1}{2n} < t < \dfrac{1}{2n} \\ 0, & t < -1/2n \end{cases}$$

Then,

(i) $d_n(t) = \dfrac{d}{dt} s_n(t) , \quad$ for $t \neq \pm 1/2n$,

(ii) $s_n(t) = \displaystyle\int_{-\infty}^{t} d_n(\tau)\,\mathrm{d}\tau,$ for all t ,

(iii) if f is any function continuous on some neighbourhood of the origin then,

$$\int_{-\infty}^{+\infty} f(t)d_n(t)\,\mathrm{d}t = n \int_{-1/2n}^{+1/2n} f(t)\,\mathrm{d}t = f(\xi_n)$$

where ξ_n is some point such that $-1/2n < \xi_n < +1/2n$ (using the First Mean Value Theorem of the Integral Calculus).

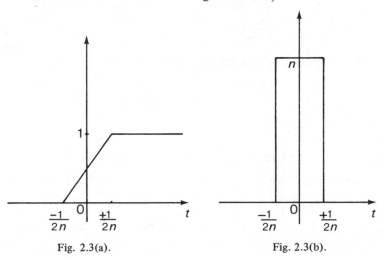

Fig. 2.3(a).   Fig. 2.3(b).

If we allow n to tend to infinity then the sequences $\{s_n\}$ and $\{d_n\}$ tend to limits u and δ in the pointwise sense:

(iv) $\displaystyle\lim_{n\to\infty} s_n(t) = u(t) = \begin{cases} 1, & t > 0 \\ 0, & t < 0 \end{cases}$; $\displaystyle\lim_{n\to\infty} d_n(t) = \delta(t) = \begin{cases} +\infty, & t = 0 \\ 0, & t \neq 0. \end{cases}$

(In actual fact, $\displaystyle\lim_{n\to\infty} s_n(t) = u_{\frac{1}{2}}(t)$ since $s_n(0) = \dfrac{1}{2}$ for all n.)

Finally, for any function f continuous on a neighbourhood of the origin we get

(v) $\displaystyle\lim_{n\to\infty} \int_{-\infty}^{+\infty} f(t)d_n(t)\,\mathrm{d}t = \lim_{n\to\infty} \int_{-1/2n}^{+1/2n} nf(t)\,\mathrm{d}t$

$\displaystyle = \lim_{n\to\infty} f(\xi_n),$ where $-\dfrac{1}{2n} < \xi_n < +\dfrac{1}{2n}$, so that, by continuity, $\displaystyle\lim_{n \to \infty} \int_{-\infty}^{+\infty} f(t) d_n(t) \, dt = f(0)$.

Thus, for $n = 1, 2, 3, \ldots$, we can construct functions d_n whose behaviour approximates more and more closely those properties which seem to be required of the delta function (treated as the formal derivative of the unit step). When we use expressions like $\displaystyle\int_{-\infty}^{+\infty} f(\tau) \delta(\tau) \, d\tau$, they may be understood simply as a convenient way of denoting the corresponding limit

$$\lim_{n \to \infty} \int_{-\infty}^{+\infty} f(\tau) d_n(\tau) \, d\tau = f(0)$$

and should not be interpreted at their face-value, namely as

$$\int_{-\infty}^{+\infty} f(\tau) \{\lim_{n \to \infty} d_n(\tau)\} \, d\tau = 0 \ .$$

2.4 THE DELTA FUNCTION AS A STIELTJES INTEGRAL

2.4.1 An entirely different way of representing the sampling operation associated with the so-called delta function is to resort to a simple generalisation of the concept of integration itself. Recall that the elementary (or Riemann) theory of the integration of bounded, continuous functions over finite intervals treats the integral as the limit of finite sums

$$\int_a^b f(t) \, dt = \lim_{\Delta t \to 0} \sum_{k=1}^{n} f(\tau_k)(t_k - t_{k-1}) \ .$$

Here the t_k are points of subdivision of the range of integration $[a,b]$, the τ_k are arbitrarily chosen points in the corresponding sub-intervals $[t_{k-1}, t_k]$, and Δt represents the largest of the quantities $\Delta_k t \equiv t_k - t_{k-1}$. In the generalisation due to Stieltjes these elements $\Delta_k t$ are replaced by quantities of the form $\Delta_k v \equiv v(t_k) - v(t_{k-1})$, where v is some fixed, monotone increasing function, and the so-called **Riemann-Stieltjes** integral appears as the limit of appropriately weighted finite sums.

Let v be a monotone increasing function and let f be any function which is continuous on the finite, closed interval $[a,b]$. By a **partition** P of $[a,b]$ we mean a subdivision of that interval by points t_k, $(0 \leqslant k \leqslant n)$, such that

$$a = t_0 < t_1 < \ldots < t_{n-1} < t_n = b \ .$$

For a given partition P let $\Delta_k v$ denote the quantity

$$\Delta_k v \equiv v(t_k) - v(t_{k-1}) , \quad k = 1, 2, \ldots, n .$$

Then, if τ_k is some arbitrarily chosen point in the sub-interval $[t_{k-1}, t_k]$, where $1 \leqslant k \leqslant n$, we can form the sum

$$\sum_{k=1}^{n} f(\tau_k) \Delta_k v . \tag{2.10}$$

It can be shown that as we take partitions of $[a,b]$ in which the points of sub-division, t_k, are chosen more and more closely together, so the corresponding sums (2.10) tend to some definite limiting value. This limit is called the **Riemann–Stieltjes** (RS) integral of f with respect to v, from $t = a$ to $t = b$, and we write

$$\int_a^b f(t) dv(t) \equiv \lim_{\Delta t \to 0} \sum_{k=1}^{n} f(\tau_k) \Delta_k v \tag{2.11}$$

where $\Delta t = \max (t_k - t_{k-1})$ for $1 \leqslant k \leqslant n$.

In the particular case in which $v(t) = t$, so that $\Delta_k v = t_{\tilde{k}} - t_{k-1}$, the RS-integral reduces to the ordinary (Riemann) integral of f over $[a,b]$:

$$\int_a^b f(t) dv(t) \equiv \int_a^b f(t) dt = \lim_{\Delta t \to 0} \sum_{k=1}^{n} f(\tau_k)(t_k - t_{k-1}).$$

More generally, if v is any monotone increasing function with a continuous derivative v', then the RS-integral of f with respect to v can always be interpreted as an ordinary Riemann integral:

$$\int_a^b f(t) dv(t) = \int_a^b f(t) v'(t) dt .$$

2.4.2 In contrast we consider a simple situation in which v is discontinuous. We shall compute the value of the Stieltjes integral

$$\int_a^b f(t) du_c(t - T)$$

where f is a function continuous on the interval $[a,b]$, T is some fixed number such that $a < T < b$, and u_c is the representative of the unit step which takes the value c (where $0 \leqslant c \leqslant 1$) at the origin. For any given partition of $[a,b]$ there will be just two possibilities:

(i) T is an interior point of some sub-interval, say $t_{k-1} < T < t_k$; in this case we will have

$$\Delta_k v = u_c(t_k - T) - u_c(t_{k-1} - T) = 1 - 0 = 1$$

while

$$\Delta_r v = 0 , \quad \text{for every } r \neq k .$$

Hence,

$$\sum_{r=1}^{n} f(\tau_r)\Delta_r v = f(\tau_k) , \quad \text{where } t_{k-1} \leqslant \tau_k \leqslant t_k \tag{2.12}$$

(ii) T is a boundary point of two adjacent sub-intervals, say $T = t_k$. This time we have

$$\Delta_k v = u_c(t_k - T) - u_c(t_{k-1} - T) = c - 0 = c ,$$

$$\Delta_{k+1} v = u_c(t_{k+1} - T) - u_c(t_k - T) = 1 - c ,$$

all other terms $\Delta_r v$ being zero. Hence,

$$\sum_{r=1}^{n} f(\tau_r)\Delta_r v = cf(\tau_k) + (1-c)f(\tau_{k+1})$$

$$= f(\tau_{k+1}) + c[f(\tau_k) - f(\tau_{k+1})] \tag{2.13}$$

where $t_{k+1} \leqslant \tau_k \leqslant T \leqslant \tau_{k+1} \leqslant t_{k+1}$.

Comparing (2.12) and (2.13), and using the fact that f is a continuous function (at least in the neighbourhood of T), it is clear that the limit to which the approximating sums tend must be $f(T)$. Note that this result is quite independent of the particular number c specified in the definition of the function u_c. (This is essentially the reason why in many contexts there is no real need to define the unit step function u at the origin.) Further, once some specific point T has been fixed, we can always choose an interval $[a,b]$ large enough to ensure that $a < T < b$. Provided only that this condition is satisfied, we have

$$\int_a^b f(t) du_c(t - T) = f(T) . \tag{2.14}$$

Since this result holds for any values of a and b such that $a < T < b$, we may take infinite limits in the integral and write the final result in a form which shows its independence of the numbers $a, b,$ and c:

If f is any function continuous on a neighbourhood of a fixed point T, and if u denotes the unit step function (left undefined at the origin) then

$$\int_{-\infty}^{+\infty} f(t)\mathrm{d}u(t-T) = f(T) . \tag{2.15}$$

In particular, if $T = 0$ then

$$\int_{-\infty}^{+\infty} f(t)\mathrm{d}u(t) = f(0) . \tag{2.16}$$

Thus the sampling property (2.9) can be legitimately expressed in terms of a Riemann–Stieltjes integral taken with respect to the unit step function u.

Expressions involving the delta function, such as $\int_{-\infty}^{+\infty} f(t)\delta(t)\mathrm{d}t$, may be regarded as conventional representations of this fact instead of being interpreted in terms of a limiting process as described in the preceding section. However, for the moment, we shall not adopt either of these approaches; as explained more fully in the next chapter, we shall concentrate instead on developing a consistent formal calculus which tells us how the symbol δ is to be used when it occurs in various algebraic contexts.

Exercises III

1. By direct evaluation of the integral $\int_{-\infty}^{+\infty} \cos t\, d_n(t)\mathrm{d}t$, where the functions d_n are as defined in Sec. 2.3, confirm that

$$\int_{-\infty}^{+\infty} \cos t\, \delta(t)\mathrm{d}t = 1,$$

in the symbolic sense attributed to integrals involving the delta function.

2. Let (g_n) be a sequence of functions defined as follows:

$$g_n(t) = \begin{cases} n^2 t + n & \text{for } -1/n \leqslant t < 0 , \\ n - n^2 t & \text{for } 0 \leqslant t < +1/n , \\ 0 & \text{for all other values of } t. \end{cases}$$

Show that, in the pointwise sense, $\lim g_n(t) = \delta(t)$, and repeat the computation of question 1, with the functions g_n in place of the functions d_n.

Sketch the graph of a typical function, G_n, defined by

$$G_n(t) = \int_{-\infty}^{t} g_n(\tau) d\tau .$$

3. Let (h_n) be a sequence of functions defined as follows:

$$h_n(t) = \begin{cases} 2n^2t & , \quad \text{for } 0 \leqslant t < 1/2n , \\ 2n - 2n^2t, & \text{for } 1/2n \leqslant t < 1/n , \\ 0 & , \quad \text{for } 1/n \leqslant t \end{cases}$$

and, $h_n(-t) = h_n(t)$.

Show that it is not the case that $\lim h_n(t) = \delta(t)$ in the pointwise sense. Repeat the computation of question 1 with the functions h_n in place of the functions d_n, and sketch the graph of a typical function, H_n, defined by

$$H_n(t) = \int_{-\infty}^{t} h_n(\tau) d\tau .$$

4.* If f is continuous on the closed interval $[a,b]$ and if $0 \leqslant c \leqslant 1$, show from first principles that

(a) $\displaystyle\int_{a}^{b} f(t) du_c(t-a) = (1-c)f(a)$

and

(b) $\displaystyle\int_{a}^{b} f(t) du_c(t-b) = cf(b)$.

2.5 HISTORICAL NOTE

Although the unit step function is usually associated with the name of Heaviside, and the delta function with that of Dirac, both concepts can be found earlier in the literature. Cauchy uses the unit step under the name "coefficient limitateur" and defines it by the formula

$$u(t) = \frac{1}{2}\left[1 + \frac{t}{\sqrt{t^2}}\right] .$$

Moreover Cauchy in 1816 (and, independently, Poisson in 1815) gave a derivation of the Fourier integral theorem by means of an argument involving what we would now recognise as a sampling operation of the type associated with the delta function. And there are similar examples of the use of what are essentially delta functions by Kirchoff, Helmholtz, and, of course, Heaviside himself. But it is undeniable that Dirac was the first to use the notation $\delta(t)$ and to state explicitly and unequivocally the more important properties which should be associated with the delta function. In "The Principles of Quantum Mechanics" (1930) Dirac refers to δ as an "improper function", and makes it quite clear that he is defining a mathematical entity of a new type, and one which cannot be identified with an ordinary function. For Dirac it is the sampling property which is the central feature of his treatment of the delta function, and he derives this by means of a formal integration by parts, as in our Sec. 2.2.2. It is fair to say that no account of the basic theory of the delta function which is both reasonably comprehensive and mathematically satisfactory was generally available until the publication in 1953 of the theory of distributions by Laurent Schwartz. (The work of Sobolev which in some respects genuinely antedates that of Schwartz did not become widely known until later.) To do justice to the ideas of Schwartz would demand a fairly extensive aquaintance on the part of the reader with the concepts and processes of modern analysis, and for the moment we shall take the subject no further. In the chapters which follow we adopt a philosophy that might well be attributed to Dirac himself. That is to say we recognise both the central importance of the so-called sampling property and that the use of the symbols $\delta(t)$ to represent that property is, at best, a purely formal convenience. Our subject is to devise systematic rules of procedure governing the use of those symbols in situations in which the normal operations of algebra and of elementary calculus may be called into play. In Chapter 7 we shall review the whole situation and attempt to give an introductory treatment of Schwartz's theory. It is to be hoped that the intervening chapters will by then have made the reader sufficently familiar with the properties and applications of delta functions to make the need for a more rigorous approach apparent.

CHAPTER 3

Properties of the Delta Function and its Derivatives

3.1 THE DELTA FUNCTION AS A GENERALISED FUNCTION

3.1.1 For each given value of the parameter c the function u_c has a well-defined pointwise specification over the entire range of values $-\infty < t < +\infty$. Now let f be an arbitrary bounded continuous function which vanishes identically outside some finite interval. Then we can write

$$\int_{-\infty}^{+\infty} f(t)u_c(t)\,dt = \int_{0}^{\infty} f(t)\,dt < +\infty . \qquad (3.1)$$

(The constraints on f are simply to ensure the existence of the integrals in (3.1) and could actually be weakened considerably.) Equation (3.1) expresses a certain operational property associated with u_c in the sense that it defines a mapping or transformation: to each continuous function f which vanishes outside some finite interval there corresponds a well-defined number, $\int_{0}^{\infty} f(t)\,dt$. Now this number is independent of the particular value $c = u_c(0)$ and hence of the particular function u_c. Recalling our convention that u is to stand for the unit step function undefined at the origin, we could equally well write (3.1) in the form.

$$\int_{-\infty}^{+\infty} f(t)u(t)\,dt = \int_{0}^{\infty} f(t)\,dt . \qquad (3.2)$$

The symbol u appearing in the integrand on the left-hand side of (3.2) may be regarded as no longer standing for any particular function but rather for an entire family of equivalent functions, u_c, any one of which would suffice to characterise the specific operation on f we have in mind. From this point of view we have no real need to specify the precise pointwise behaviour of the unit step; it is enough that the symbol u appearing in (3.2) is known to define a certain mapping of continuous functions f into numbers:

$$f \rightarrow \int_0^{\infty} f(t)\,\mathrm{d}t \ .$$

In the same way the one important feature about the so-called delta function, $\delta(t)$, is that it affords a convenient means of representing an operation, defined at least for all functions continuous on a neighbourhood of the origin, which maps or transforms each such function f into the value $f(0)$ which it assumes at the origin. We use the familiar integral notation

$$f \rightarrow f(0) = \int_{-\infty}^{+\infty} f(t)\delta(t)\,\mathrm{d}t \qquad (3.3)$$

as a symbolic representation of this fact, but make no use of the apparent point-wise behaviour of $\delta(t)$ which the notation of (3.3) suggests.

3.1.2 Symbols such as u and δ when regarded as specifying operations on certain classes of functions (rather than as standing for functions $u(t)$ and $\delta(t)$ in their own right) are usually referred to as **generalised functions**. In what follows we shall develop a set of rules which allow us to interpret expressions involving u and δ (and other, allied, symbols denoting generalised functions) and to apply to them the usual processes of elementary calculus. Throughout it is only the operational significance of u and δ which will concern us. No inferences can safely be drawn from the apparent pointwise behaviour of $\delta(t)$ although, on occasion, such inferences may turn out to be correct. At the same time it is often helpful to think of $\delta(t)$ as though it denoted a function in the ordinary sense which takes the value zero everywhere except at the origin where its value must be presumed infinite. Provided that this pointwise specification is treated as purely descriptive and is not used to justify algebraic manipulations etc., the abuse of notation involved is harmless (and, indeed, too well established to be wholly ignored).

3.1.3 More generally, consider the expression $\delta(t-a)$ where a is any fixed real number. Formally we have

$$\int_{-\infty}^{+\infty} f(t)\delta(t-a)\,\mathrm{d}t = \int_{-\infty}^{+\infty} f(\tau+a)\delta(\tau)\,\mathrm{d}\tau = \left[f(\tau+a)\right]_{\tau=0} = f(a) \ .$$

In general we use the symbol f_a to denote the **translate of the function** f **with respect to** a:

$$f_a(t) \equiv f(t-a) \ . \qquad (3.4)$$

Accordingly the calculation performed above suggests that the translate of the delta function, δ_a, represents, or is characterised by, that operation which carries any function f, continuous on a neighbourhood of a, into the value $f(a)$ which that function assumes at a. Moreover the evaluation of a Stieltjes integral of the form

$$\int_{-\infty}^{+\infty} f(t)\mathrm{d}u(t-a)$$

(as in Sec. 2.4) is enough to show that δ_a may be regarded as the formal derivative of the translated step function $u(t-a)$.[†]

Hence we take as our starting point the generalised function δ_a which is defined by the sampling operation

$$f \to f(a)$$

where f is any function which is continuous at least on some neighbourhood of a. Symbolically we write

$$\int_{-\infty}^{+\infty} f(t)\delta_a(t)\mathrm{d}t \equiv \int_{-\infty}^{+\infty} f(t)\delta(t-a)\mathrm{d}t = f(a)$$

and,

$$\frac{\mathrm{d}}{\mathrm{d}t}u(t-a) = \delta(t-a) \tag{3.5}$$

3.2 ADDITION AND MULTIPLICATION

3.2.1 If h_1 and h_2 are any two ordinary functions then their **pointwise sum** is defined as the function h_3 whose value at each point t is the sum of the values assumed by h_1 and h_2 at that point:

$$h_3(t) \equiv (h_1 + h_2)(t) = h_1(t) + h_2(t) \tag{3.6}$$

For any other function f for which all the integrals involved exist we have

$$\int_{-\infty}^{+\infty} f(t)h_3(t)\mathrm{d}t = \int_{-\infty}^{+\infty} f(t)\{h_1(t) + h_2(t)\}\mathrm{d}t$$

$$= \int_{-\infty}^{+\infty} f(t)h_1(t)\mathrm{d}t + \int_{-\infty}^{+\infty} f(t)h_2(t)\mathrm{d}t . \tag{3.7}$$

Where delta functions are concerned we cannot invoke pointwise behaviour, so that (3.6) is of no relevance in this context. However, we can generalise (3.7) in

† In view of the special significance of the subscript in $u_c(t)$, as described in Sec. 2.1.2, we shall always denote a translate of the unit step function by writing it in full, as here.

an obvious way so as to attach a meaning to combinations like $h + \delta$, where h is an ordinary function, and to expressions of the form $\delta_a + \delta_b$. For any function f, continuous on a neighbourhood of the origin, we may write

$$\int_{-\infty}^{+\infty} f(t)\{h(t) + \delta(t)\}dt = \int_{-\infty}^{+\infty} f(t)h(t)dt + \int_{-\infty}^{+\infty} f(t)\delta(t)dt$$

$$= \int_{-\infty}^{+\infty} f(t)h(t)dt + f(0) . \tag{3.8}$$

Similarly, provided that f is continuous on a neighbourhood of a and on a neighbourhood of b,

$$\int_{-\infty}^{+\infty} f(t)\{\delta_a(t) + \delta_b(t)\}dt = \int_{-\infty}^{+\infty} f(t)\delta_a(t)dt + \int_{-\infty}^{+\infty} f(t)\delta_b(t)dt$$

$$= f(a) + f(b) \tag{3.9}$$

Equations (3.8) and (3.9) effectively define sampling properties which characterise $h + \delta$ and $\delta_a + \delta_b$ respectively as generalised functions. These results clearly depend on the assumption that the distributive law of ordinary algebra should remain valid in expressions involving delta functions and ordinary functions.

3.2.2 Now consider the formal product $\phi\delta$, where ϕ is a function continuous at least on some neighbourhood of the origin. If f is any other such function, and we assume that the associative law of multiplication continues to hold, then we should have

$$\int_{-\infty}^{+\infty} f(t)\{\phi(t)\,\delta(t)\}dt = \int_{-\infty}^{+\infty} \{f(t)\phi(t)\}\delta(t)dt = f(0)\phi(0) . \tag{3.10}$$

In particular if ϕ is a constant function, say $\phi(t) = k$ for all t, then we get

$$\int_{-\infty}^{+\infty} f(t)\{k\delta(t)\}dt = \int_{-\infty}^{+\infty} \{kf(t)\}\delta(t)dt$$

$$= k\int_{-\infty}^{+\infty} f(t)\delta(t)dt = kf(0) . \tag{3.11}$$

Note that we may write the result (3.10) in the equivalent form

$$\phi(t)\delta(t) = \phi(0)\delta(t) \tag{3.12}$$

in the sense that the formal product $\phi(t)\delta(t)$ has the same operational significance as that given to $k\delta(t)$ by equation (3.11) when the constant k has the value $\phi(0)$.

3.2.3 Remark[*]

In considering products involving delta functions we have been careful to confine attention to *continuous* multipliers; this is because the characteristic sampling property of the delta function is itself only defined for continuous functions (or, at most, for those functions continuous on some neighbourhood of the origin). One way of removing this restriction is suggested by the following considerations.

Suppose that f_0 is some fixed, continuous, function such that $f_0(0) = 1$. For an arbitrary continuous function f let $\hat{f}$ denote the function defined by

$$\hat{f}(t) = f(t) - f(0)f_0(t) \ .$$

Since $\hat{f}$ is clearly continuous and such that $\hat{f}(0) = 0$, it follows that the product $u_c(t)\hat{f}(t)$ is defined for all t and is continuous everywhere and independent of the number c. Hence we may write

$$\int_{-\infty}^{+\infty} \hat{f}(t)\{u(t)\delta(t)\}\,dt \equiv \int_{-\infty}^{+\infty} \{\hat{f}(t)u_c(t)\}\delta(t)\,dt = \hat{f}(0)c = 0 \ .$$

Once again assuming that all the products involved obey the usual associative and distributive laws this gives

$$0 = \int_{-\infty}^{+\infty} \{f(t) - f(0)f_0(t)\}\{u(t)\delta(t)\}\,dt$$

$$= \int_{-\infty}^{+\infty} f(t)\{u(t)\delta(t)\}\,dt - f(0)\int_{-\infty}^{+\infty} f_0(t)\{u(t)\delta(t)\}\,dt$$

that is

$$\int_{-\infty}^{+\infty} f(t)\{u(t)\delta(t)\}\,dt = f(0)\int_{-\infty}^{+\infty} f_0(t)\{u(t)\delta(t)\}\,dt \ . \qquad (3.13)$$

Thus we could define the sampling operation characteristic of the formal product $u\delta$ for continuous functions f provided only that it is well defined in respect of one particular function f_0. In fact from (3.13) we would conclude that

$$u(t)\delta(t) = k\delta(t) \qquad (3.14)$$

where k is an arbitrary constant.

There are some arguments which suggest that the most appropriate value for

k in (3.14) is $1/2$. For example a formal differentiation of the function $u^2(t)$ gives

$$\frac{d}{dt}u^2(t) = 2u(t)\frac{d}{dt}u(t) = 2u(t)\delta(t) = 2k\delta(t)$$

But, clearly, we have $u^2(t) = u(t)$ so that $\dfrac{d}{dt}u^2(t) = \dfrac{d}{dt}u(t) = \delta(t)$. However, there are contexts in which the choice $k = 1/2$ is not as convenient as might be supposed from the foregoing. We shall consider this point further in Chapter 5 in connection with the problem of defining the Laplace Transform of the delta function.

Exercises I

1. Find formal simplified equivalent expressions for:

   (a) $\cos t + \sin t\, \delta(t)$;   (b) $\sin t + \cos t\, \delta(t)$;   (c) $1 + 2e^t\delta(t-1)$.

2. Evaluate the following integrals:

   (a) $\displaystyle\int_{-\infty}^{+\infty} (t^2 + 3t + 5)\delta(t)\,dt$;   (b) $\displaystyle\int_{-\infty}^{+\infty} \frac{\cos x\, \delta(x)}{2e^x + 1}\,dx$;

   (c) $\displaystyle\int_{-\infty}^{+\infty} e^{-t^2}\delta(t-2)\,dt$;   (d) $\displaystyle\int_{-\infty}^{+\infty} \sinh 2t\, \delta(2-t)\,dt$;

   (e) $\displaystyle\int_{-\infty}^{+\infty} \frac{1 + e^{2x}\delta(x+1)}{1+x^2}\,dx$;   (f) $\displaystyle\int_{-\infty}^{+\infty} t^2 \sum_{k=1}^{n} \delta(t-k)\,dt$;

   (g) $\displaystyle\int_{-\infty}^{+\infty} \{e^{2\theta}u(-\theta) - \cos 2\theta\, \delta(\theta) + e^{-2\theta}u(\theta)\}\,d\theta$;

   (h) $\displaystyle\int_{-\infty}^{+\infty} \{e^{\tau}\delta_\pi(\tau) + e^{-\tau}\delta_{-\pi}(\tau)\}\,d\tau$.

3.3 DIFFERENTIATION

3.3.1 We can now attach meaning to the term "derivative" in the case of a function which has one or more jump discontinuities. Let ϕ_1 and ϕ_2 be continuously differentiable functions and let f be defined as the function which is equal to ϕ_1 for all $t < a$ and equal to ϕ_2 for all $t > a$:

$$f(t) = \phi_2(t)u(t-a) + \phi_1(t)u(a-t) .$$

Then,

$$\frac{d}{dt}f(t) = u(a-t)\frac{d}{dt}\phi_1(t) + \phi_1(t)\frac{d}{dt}u(a-t)$$

$$+ u(t-a)\frac{d}{dt}\phi_2(t) + \phi_2(t)\frac{d}{dt}u(t-a).$$

Now,

$$\frac{d}{dt}u(t-a) = \frac{du(t-a)}{d(t-a)}\frac{d(t-a)}{dt} = \delta(t-a) \equiv \delta_a(t)$$

and

$$\frac{d}{dt}u(a-t) = \frac{du(a-t)}{d(a-t)}\frac{d(a-t)}{dt} = -\delta(a-t) = -\delta_a(t).$$

Hence,

$$\frac{d}{dt}f(t) = \phi_1'(t)u(a-t) + \phi_2'(t)u(t-a) - \phi_1(t)\delta_a(t) + \phi_2(t)\delta_a(t)$$

$$= \phi_1'(t)u(a-t) + \phi_2'(t)u(t-a) + \{\phi_2(a) - \phi_1(a)\}\delta_a(t). \quad (3.15)$$

The discontinuity at $t = a$ thus gives rise to a delta function at $t = a$ multiplied by the saltus of the function at that point, namely by the number $k = f(a+) - f(a-)$. It is customary to refer to this as a delta function of **strength** k at $t = a$. Integration of (3.15) will reintroduce the jump discontinuity in the primitive, f.

In particular, suppose that $\phi_1(t) = \phi(t)$ and that $\phi_2(t) = \phi(t) + k$. Then (3.15) becomes

$$\frac{d}{dt}f(t) = \phi'(t)u(a-t) + \phi'(t)u(t-a) + \{(\phi(a) + k) - \phi(a)\}\delta_a(t)$$

$$= \phi'(t) + k\delta_a(t). \quad (3.16)$$

Further, if x and t are any two numbers such that $x < a < t$ then we get

$$\int_x^t \frac{d}{d\tau}f(\tau)d\tau = \int_x^a \phi'(\tau)d\tau + \int_a^t \phi'(\tau)d\tau + k\int_x^t \delta(\tau - a)d\tau$$

$$= \phi(a) - \phi(x) + \phi(t) - \phi(a) + k = f(t) - f(x),$$

the integral involving the delta function at a being well-defined just because $x < a < t$. (See Sec. 3.6.2 below.)

3.3.2 This extension of the concept of differentiation does give rise to certain notational difficulties. It is standard practice to denote by f' the derivative of a function f even when that derivative is not defined everywhere. For example, let $f(t) = |t|$, and write $\phi(t) = \text{sgn } t$. Then f is differentiable everywhere except at the origin, and we have $f'(t) = \phi(t)$, for $t \neq 0$. Further, no matter what value we assign to $\phi(0)$, it is always the case that

$$\int_{-1}^{t} \phi(\tau)\,d\tau = \int_{-1}^{t} f'(\tau)\,d\tau = |t| - 1 = f(t) - f(-1).$$

Now consider the function ϕ itself. This is again a function differentiable (in the usual sense) everywhere except at the origin. If we adopt the usual conventions of elementary calculus then we would write $\phi'(t) = 0$, for all $t \neq 0$. Moreover, with the ordinary interpretation of the integration sign, it would be entirely correct to write

$$\int_{-1}^{t} \phi'(\tau)\,d\tau = \int_{-1}^{t} 0\,d\tau = 0.$$

However, we now wish "derivative" in this case to stand for $2\delta(t)$ so that a formal integration will recover the original discontinuous function, sgn t, at least up to an arbitrary constant. Some mode of distinguishing between the elementary concept of derivative and the generalised sense which may include delta functions seems called for. Usually the context will make clear which is the intended meaning. When there is danger of real confusion on this point we can always use the following convention:

if $\qquad\qquad f(t) = \phi_1(t)u(a - t) + \phi_2(t)u(t - a)$

as above, then $\quad f'(t) = \phi_1'(t)u(a - t) + \phi_2'(t)u(t - a) \quad$ for $t \neq a$,

but $\qquad Df(t) = \phi_1'(t)u(a - t) + \phi_2'(t)u(t - a) + \delta_a(t)\{\phi_2(a) - \phi_1(a)\}$

that is $\qquad\qquad Df(t) \equiv f'(t) + \{f(a+) - f(a-)\}\delta(t - a).$ $\qquad$ (3.17)

We shall call f' the **classical** derivative and Df the **generalised** derivative of f.

Exercises II

1. Find the first derivative of:

    (a) $u(1 + t) + u(1 - t)$;   (b) $[1 - u(t)]\cos t$;

    (c) $\left[u\left(t - \dfrac{\pi}{2}\right) - u\left(t - \dfrac{3\pi}{2}\right)\right]\sin t$;

    (d) $\{u(t + 2) + u(t) - u(t - 2)\}e^{-2t}$.

2. Find the first and second derivatives of:

    (a) $|t|$;   (b) $e^{-|t|}$;   (c) $\sin |t|$.

3.* Show that if $f(t) = \tanh 1/t$, for $t \neq 0$, then f has a jump discontinuity at the origin; find the (generalised) first derivative of f. Show also that if we write

$$g(t) = \frac{d}{dt} \tanh 1/t \;, \quad \text{for } t \neq 0 \;, \quad \text{and } g(0) = 0 \;,$$

then g is a function which is continuous for all t.

3.4 DERIVATIVES OF THE DELTA FUNCTION

3.4.1 Suppose now that f is a function which is continuously differentiable in some neighbourhood of the origin. Then, for any number $a \neq 0$ which is sufficiently small in absolute magnitude, we would have:

$$\int_{-\infty}^{+\infty} f(t) \left[\frac{\delta(t+a) - \delta(t)}{a} \right] dt = \frac{1}{a} \left[\int_{-\infty}^{+\infty} f(t)\delta(t+a)\,dt - \int_{-\infty}^{+\infty} f(t)\delta(t)\,dt \right].$$

$$= \frac{1}{a} \left[\int_{-\infty}^{+\infty} f(\tau - a)\delta(\tau)\,d\tau - f(0) \right] = \frac{1}{a} \{ f(-a) - f(0) \} \;.$$

Hence,

$$\lim_{a \to 0} \int_{-\infty}^{+\infty} f(t) \left\{ \frac{\delta(t+a) - \delta(t)}{a} \right\} dt = \lim_{a \to 0} \frac{f(-a) - f(0)}{a} = -f'(0) \;.$$

This result suggests that a generalised function δ', designed to play the role of derivative of the delta function itself, should be characterised by the following sampling property:

If f is any function which has a continuous derivative f' at least in some neighbourhood of the origin, then

$$\int_{-\infty}^{+\infty} f(\tau)\delta'(\tau)\,d\tau = -f'(0) \;. \tag{3.18}$$

This is precisely what might have been inferred from a formal integration by parts:

$$\int_{-\infty}^{+\infty} f(t)\delta'(t)\,dt = \left[f(t)\delta(t) \right]_{-\infty}^{+\infty} - \int_{-\infty}^{+\infty} f'(t)\delta(t)\,dt =$$

$$\left[f(0)\delta(t) \right]_{-\infty}^{+\infty} - f'(0) = -f'(0)$$

(Note that strictly this last step does involve a tacit appeal to the pointwise behaviour of the delta function. However, all that we are really saying is that the operational property of δ is confined absolutely to the point $t = 0$. At any other points it has no effect, and this is what is really meant by saying that it "evaluates to zero" at such points.)

A straightforward generalisation of the above considerations leads to the following definition:

For any given positive integer n the generalised function $\delta^{(n)}$, (the nth derivative of the delta function), is defined by the characteristic sampling property

$$\int_{-\infty}^{+\infty} f(\tau)\delta^{(n)}(\tau)\,d\tau = (-1)^n f^{(n)}(0) \tag{3.19}$$

where f is any function with continuous derivatives at least up to the nth order in some neighbourhood of the origin.

3.4.2 For the most part the properties of the derivatives $\delta^{(n)}$ are fairly obvious generalisations of the corresponding properties of the delta function itself, and can be established by similar arguments. We list them briefly here, enlarging on those of particular difficulty and importance.

(i) *Translation* For any continuously differentiable function f and any scalar a we have

$$\int_{-\infty}^{+\infty} f(\tau)\delta_a'(\tau)\,d\tau \equiv \int_{-\infty}^{+\infty} f(\tau)\delta'(\tau - a)\,d\tau$$

$$= \int_{-\infty}^{+\infty} f(t + a)\delta'(t)\,dt = \left[-f'(t + a)\right]_{t=0} = -f'(a) . \tag{3.20}$$

More generally,

$$\int_{-\infty}^{+\infty} f(\tau)\delta_a^{(n)}(\tau)\,d\tau \equiv \int_{-\infty}^{+\infty} f(\tau)\delta^{(n)}(\tau - a)\,d\tau = (-1)^n f^{(n)}(a) \tag{3.21}$$

for any function f which has continuous derivatives at least up to the nth order in some neighbourhood of the point $t = a$.

(ii) *Addition* Let n and m be positive integers. If f is any function which has continuous derivatives at least up to the nth order on a neighbourhood of the point $t = a$, and at least up to the mth order on a neighbourhood of the point $t = b$, then

$$\int_{-\infty}^{+\infty} f(\tau)\{\delta_a^{(n)}(\tau) + \delta_b^{(m)}(\tau)\}\,d\tau = (-1)^n f^{(n)}(a) + (-1)^m f^{(m)}(b) \quad (3.22)$$

(iii) *Scalar multiplication* The product $k\delta^{(n)}$, where k is some fixed number, is defined in terms of the following sampling operation:

$$\int_{-\infty}^{+\infty} f(\tau)\{k\delta^{(n)}(\tau)\}\,d\tau = \int_{-\infty}^{+\infty} \{f(\tau)k\}\delta^{(n)}(\tau)\,d\tau = (-1)^n k f^{(n)}(0) \quad (3.23)$$

In particular if $k = 0$ then we may write $k\delta^{(n)}(t) = 0$.

(iv) *Multiplication by a function* If ϕ is a fixed function which is continuously differentiable in some neighbourhood of the origin, then a meaning can be given to the product $\phi\delta'$:

$$\int_{-\infty}^{+\infty} f(\tau)\{\phi(\tau)\delta'(\tau)\}\,d\tau = \int_{-\infty}^{+\infty} \{f(\tau)\phi(\tau)\}\delta'(\tau)\,d\tau$$

$$= \left[-\frac{d}{d\tau}\{f(\tau)\phi(\tau)\}\right]_{\tau=0} = -f'(0)\phi(0) - f(0)\phi'(0) . \quad (3.24)$$

Thus, assuming associativity for the products involved, the expression $\phi\delta'$ is seen to be formally equivalent to the linear combination

$$a\delta'(t) + b\delta(t)$$

where $a = \phi(0)$ and $b = -\phi'(0)$.

Similarly, if ϕ is some fixed function which is twice continuously differentiable in a neighbourhood of the origin then we would have

$$\int_{-\infty}^{+\infty} f(\tau)\{\phi(\tau)\delta''(\tau)\}\,d\tau = \int_{-\infty}^{+\infty} \{f(\tau)\phi(\tau)\}\delta''(\tau)\,d\tau$$

$$= \left[+\frac{d^2}{d\tau^2}\{f(\tau)\phi(\tau)\}\right]_{\tau=0} = f''(0)\phi(0) + 2f'(0)\phi'(0) + f(0)\phi''(0)$$

so that the product $\phi\delta''$ is equivalent to the linear combination

$$\phi(0)\delta''(t) - 2\phi'(0)\delta'(t) + \phi''(0)\delta(t) . \quad (3.25)$$

In general the product $\phi\delta^{(n)}$, where ϕ has continuous derivatives at least up to the nth order on a neighbourhood of the origin, reduces to

$$\phi(0)\delta^{(n)}(t) - n\phi'(0)\delta^{(n-1)}(t) + \frac{n(n-1)}{2!}\phi''(0)\delta^{(n-2)}(t) - \ldots\ldots$$

$$\ldots\ldots + (-1)^n \phi^{(n)}(0)\delta(t) \quad (3.26)$$

Example The expressions $\sin t\,\delta'(t)$ and $\cos t\,\delta'(t)$ may be replaced by simpler equivalent expressions as follows:

$$\sin t\,\delta'(t) = (\sin 0)\delta'(t) - (\cos 0)\delta(t) = -\delta(t)$$

$$\cos t\,\delta'(t) = (\cos 0)\delta'(t) - (-\sin 0)\delta(t) = \delta'(t)\,.$$

That this is the case can be seen by evaluating typical integrals involving these products. If f is any continuously differentiable function then we have

$$\int_{-\infty}^{+\infty} f(\tau)\{\sin \tau\,\delta'(\tau)\}\mathrm{d}\tau = \int_{-\infty}^{+\infty} \{f(\tau)\sin \tau\}\delta'(\tau)\mathrm{d}\tau$$

$$= \left[-\frac{\mathrm{d}}{\mathrm{d}\tau}\{f(\tau)\sin \tau\}\right]_{\tau=0} = -f(0) = -\int_{-\infty}^{+\infty} f(\tau)\delta(\tau)\mathrm{d}\tau$$

and

$$\int_{-\infty}^{+\infty} f(\tau)\{\cos \tau\,\delta'(\tau)\}\mathrm{d}\tau = \int_{-\infty}^{+\infty} \{f(\tau)\cos \tau\}\delta'(\tau)\mathrm{d}\tau$$

$$= \left[-\frac{\mathrm{d}}{\mathrm{d}\tau}\{f(\tau)\cos \tau\}\right]_{\tau=0} = -f'(0) = \int_{-\infty}^{+\infty} f(\tau)\delta'(\tau)\mathrm{d}\tau\,.$$

3.5 POINTWISE DESCRIPTION OF $\delta'(t)$

3.5.1 In deriving the foregoing results, no appeal has been made to any supposed pointwise behaviour of the derivatives δ', δ'', ..., etc. As in the case of the delta function itself no such appeal is actually necessary and it is usually better avoided. Moreover, as the following considerations show, the behaviour of even the first derivative, δ', at the origin is extremely difficult to describe in terms of ordinary functions. Take the sequence $\{d_n\}$ of functions chosen in Sec. 2.3 of Chapter 2 to exhibit the delta function, δ, as a limit. For each n the derivative d_n' consists of a pair of delta functions located at the points $-\tfrac{1}{2}n$, $+\tfrac{1}{2}n$. These are shown in Fig. 3.1 as spikes in accordance with a fairly well-established convention.

As n increases the delta functions increase in magnitude and approach nearer and nearer to the origin from either side; in the limit we apparently have to deal with coincident delta functions of opposite sign and of arbitrarily large magnitude. Operationally, on the other hand, the significance of the sequence $\{d_n'\}$ is easy to interpret. If f is any function which is continuously differentiable on a neighbourhood of $t = 0$ then, for all sufficiently large values of n,

$$\int_{-\infty}^{+\infty} f(\tau)d'_n(\tau)\,d\tau = \int_{-\infty}^{+\infty} f(\tau)n\left[\delta\left(\tau+\frac{1}{2n}\right)-\delta\left(\tau-\frac{1}{2n}\right)\right]d\tau$$

$$= n\int_{-\infty}^{+\infty} f(\tau)\delta\left(\tau+\frac{1}{2n}\right)d\tau - n\int_{-\infty}^{+\infty} f(\tau)\delta\left(\tau-\frac{1}{2n}\right)d\tau$$

$$= \frac{f(-1/2n)-f(1/2n)}{1/n}$$

and, as n goes to infinity, this tends to the limit $-f'(0)$.

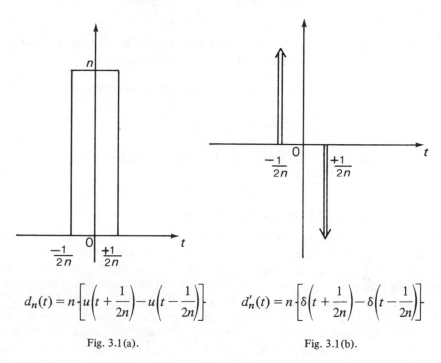

$$d_n(t) = n\left\{u\left(t+\frac{1}{2n}\right)-u\left(t-\frac{1}{2n}\right)\right\}$$ 　　　　 $$d'_n(t) = n\left[\delta\left(t+\frac{1}{2n}\right)-\delta\left(t-\frac{1}{2n}\right)\right]$$

　　　　　Fig. 3.1(a). 　　　　　　　　　　　　　　　Fig. 3.1(b).

3.5.2 The same point can be made, somewhat less dramatically, by using a sequence of ordinary functions which converges (in some sense) to δ'. For example if we write

$$p_n(t) = \frac{n^2}{4}\left[u\left(t+\frac{2}{n}\right)-2u(t)+u\left(t-\frac{2}{n}\right)\right]$$

then δ' can be taken as $\lim p_n(t)$. In the neighbourhood of $t = 0$, as $n \to \infty$, the functions p_n assume arbitrarily large values both positive and negative.

Exercises III

1. Evaluate the following integrals:

(a) $\displaystyle\int_{-\infty}^{+\infty} e^{at}\sin bt\, \delta^{(n)}(t)\,dt\,,\quad$ for $n=1,2,3$;

(b) $\displaystyle\int_{-\infty}^{+\infty}\frac{\delta(t)\cos t - \delta'(t)\cos 2t}{1+t^2}\,dt\,$;

(c) $\displaystyle\int_{-\infty}^{+\infty}\{t^3+2t+3\}\{\delta'(t-1)+2\delta''(t-2)\}\,dt\,.$

2. Show that δ' behaves like an *odd* function, that is, that the expression $\delta'(-t)$ is formally equivalent to the expression $-\delta'(t)$.

3.* Show that if ϕ is a twice continuously differentiable function then,

$$\frac{d}{dt}\{\phi(t)\delta'(t)\} = \phi'(t)\delta'(t) + \phi(t)\delta''(t)$$

in the sense that both sides have the same operational significance.

3.6 INTEGRATION OF DELTA FUNCTIONS

3.6.1 The significance of the delta function (and of each of its derivatives) is intimately bound up with a certain conventional use of the notation for integration. This usage is consistent with the familiar view of integration as a process which is, in some sense, inverse to that of differentiation. Thus, recall the statement made in Sec. 3.1.1 that the delta function, δ, itself is to be regarded as the derivative of the unit step function u. As already remarked in Sec. 2.2.3 this statement is to be interpreted in the following sense. If $t\neq 0$ the function $u(t-\tau)$ is a continuous function of τ in a neighbourhood of $\tau=0$; accordingly the formal product $u(t-\tau)\delta(\tau)$ is meaningful and we may write

$$\int_{-\infty}^{t}\delta(\tau)\,d\tau = \int_{-\infty}^{+\infty} u(t-\tau)\delta(\tau)\,d\tau = u(t)\,. \qquad (3.27)$$

In particular this gives

$$\int_{-\infty}^{+\infty}\delta(\tau)\,d\tau = 1\,. \qquad (3.28)$$

More generally, recall that we define $\delta^{(n)}$ as the derivative of $\delta^{(n-1)}$:

$$\delta^{(n)}(t) = \frac{d}{dt}\,\delta^{(n-1)}(t)\,.$$

Since (for $t \neq 0$) the function $u(t-\tau)$ is actually differentiable to all orders as a function of τ in a neighbourhood of $\tau = 0$ we may write

$$\int_{-\infty}^{t} \delta^{(n)}(\tau)\,d\tau = \int_{-\infty}^{+\infty} u(t-\tau)\delta^{(n)}(\tau)\,d\tau = \delta^{(n-1)}(t)\,, \quad n \geqslant 1\,. \quad (3.30)$$

without loss of consistency.

Finally, since the function $f(t) \equiv 1$ is, trivially, infinitely differentiable we have the result

$$\int_{-\infty}^{+\infty} \delta^{(n)}(\tau)\,d\tau = 0 \quad \text{for } n \geqslant 1\,. \quad (3.30)$$

3.6.2 *Finite limits* In general for definite integrals involving one or more finite limits of integration the presence of delta functions (or derivatives of delta functions) in the integrand can present problems of interpretation. The fundamental sampling property which characterises, say, $\delta^{(n)}$ presupposes continuity of the nth derivative of the integrand at least on a neighbourhood of the origin. If the range of integration includes the origin as an interior point (i.e. if the integral is taken over $[a,b]$, where $a < 0 < b$) then there is no difficulty; the function $u(t-a) - u(t-b)$ is certainly continuous and differentiable to all orders on a neighbourhood of the origin and we have

$$\int_{a}^{b} f(\tau)\delta^{(n)}(\tau)\,d\tau \equiv \int_{-\infty}^{+\infty} f(\tau)\{u(t-a) - u(t-b)\}\delta^{(n)}(\tau)\,d\tau$$

$$= (-1)^n f^{(n)}(0)\,. \quad (n \geqslant 0)\,. \quad (3.21)$$

Again, if the origin lies wholly outside the interval $[a,b]$ (so that $0 < a$, or $b < 0$) then no problem arises; trivially we have

$$\int_{a}^{b} f(\tau)\delta^{(n)}(\tau)\,d\tau = \int_{-\infty}^{+\infty} f(\tau)\{u(t-a) - u(t-b)\}\delta^{(n)}(\tau)\,d\tau = 0\,. \quad (3.32)$$

(All these remarks obviously apply to any integrand containing a factor of the form $\delta_c^{(n)}(t)$ if we replace "the origin" by "the point c" throughout.)

The trouble starts whenever the origin coincides with one or other of the limits of integration. Specifically, consider the case of the expression

$$\int_0^\infty \delta(\tau)d\tau .\tag{3.33}$$

As it stands this is completely undefined, and we appear to be free to assign to it any value which turns out to be convenient. The discussion in Section 3.2.3 shows that there are strong reasons for choosing the value $\frac{1}{2}$, but the grounds for this choice are rather more complicated than may at first appear, and we defer detailed consideration until a later stage. For the moment we merely note that the problem is not an isolated one. If we write

$$\int_0^\infty \delta(\tau)d\tau = \int_{-\infty}^{+\infty} u(t)\delta(t)dt$$

then assigning a suitable value to (3.33) is seen to be linked with at least two other problems, namely:

(i) extension of the sampling property to apply to discontinuous integrands,
(ii) the meaning of formal products like $\phi\delta$, when ϕ is not continuous at $t = 0$.

3.7 CHANGE OF VARIABLE

3.7.1 Bearing the remarks of the preceding section in mind it remains to consider the problem of change of variable where integrals involving delta functions are concerned; that is, we need to be able to interpret expressions like $\delta^{(n)}\{\phi(t)\}$ when they occur in an integrand. Recall first that, for ordinary, sufficiently well-behaved functions, the substitution $x = \phi(t)$ leads to an equivalence of the form

$$\int_a^b f(t)h\{\phi(t)\}dt = \int_{\phi(a)}^{\phi(b)} f\{\phi^{-1}(x)\}h(x)\{\phi^{-1}(x)\}'dx\tag{3.34}$$

where, for $\phi(a) \leqslant x \leqslant \phi(b)$, we assume that $\phi^{-1}(x)$ is uniquely defined and has a uniquely defined derivative, $[\phi^{-1}(x)]'$. By imposing certain restrictions on $\phi(t)$ we can establish the validity of (3.34) when $h\{..\}$ is replaced by $\delta\{..\}$, $\delta'\{..\}$, and so on.

However, it is often the case that a simple heuristic argument is enough to reduce the integrals concerned to an equivalent form which is easy to evaluate. In general when $x = \phi(t)$ we can write

$$\delta(x) = \frac{du}{dx}(x) = \frac{d}{dt}u\{\phi(t)\}\frac{dt}{dx} = \frac{d}{dt}u\{\phi(t)\}\bigg/\frac{d\phi}{dt}$$

and this last expression is easy to interpret whenever $u\{\phi(t)\}$ itself reduces to a simpler form. Granted this much the method readily extends to delta functions of higher order, since we have

$$\delta'(x) = \frac{d}{dx}\,\delta(x) = \frac{d}{dt}\,\delta\{\phi(t)\}\,\frac{dt}{dx} = \frac{d}{dt}\,\delta\{\phi(t)\}\Big/\left|\frac{d\phi}{dt}\right|$$

and so on. The process is most easily explained in terms of one or two explicit examples as discussed below.

3.7.2 Suppose first that $\phi(t)$ increases monotonely from $\phi(a)$ to $\phi(b)$ as t goes from a to b; suppose also that $\phi(c) = 0$ for some point c such that $a < c < b$ and that $\phi(a) < \phi(c) < \phi(b)$. Using the fact that

$$\frac{d}{dx}[\phi^{-1}(x)] = 1\Big/\frac{d}{dt}[\phi(t)]$$

we have

$$\int_a^b f(t)\delta\{\phi(t)\}dt = \int_{\phi(a)}^{\phi(b)} f\{\phi^{-1}(x)\}\delta(x)\{\phi^{-1}(x)\}'dx$$

$$= f(c)\,[\phi^{-1}(x)]'_{x=\phi(c)} = \frac{1}{\phi'(c)}f(c)\,.$$

If $\phi(t)$ happens to decrease monotonely as t goes from a to b then the same result is obtained apart from a change of sign. Since in this case we would have $\phi'(c) < 0$ the two results can be summed up in the single formula:

if $\phi(t)$ monotone, with $\phi(c) = 0$ and $\phi'(c) \neq 0$, then

$$\delta\{\phi(t)\} = \frac{1}{|\phi'(c)|}\delta_c(t)\,. \tag{3.35}$$

In particular,

$$\delta\{\alpha t - \beta\} = \frac{1}{|\alpha|}\delta\!\left(t - \frac{\beta}{\alpha}\right). \tag{3.36}$$

In point of fact it is simpler to derive (3.35) by noting that under the particular hypotheses on ϕ we must have the equivalence

$$u\{\phi(t)\} = u(t - c)\,, \quad \text{for } \phi \text{ monotone increasing,}$$

$$u\{\phi(t)\} = u(c - t)\,, \quad \text{for } \phi \text{ monotone decreasing.}$$

Since $\dfrac{d}{dt}u(t - c) = \delta(t - c)$, and $\dfrac{d}{dt}u(c - t) = -\delta(t - c)$, (3.35) follows immediately. A similar argument can then be applied to $\delta'\{\phi(t)\}$ and so on.

Examples

(i) Let $I_1 = \int_0^2 e^{4t}\delta(2t-3)dt$, so that $x = 2t - 3$, $t = \frac{1}{2}(x + 3)$, and $dt = \frac{1}{2}dx$.

Then,

$$I_1 = \int_{-3}^1 e^{2(x+3)}\delta(x)\frac{1}{2}dx = \frac{1}{2}e^6.$$

Alternatively, note that $u(2t - 3) = u(t - 3/2)$; hence

$$\frac{d}{dt}u(2t - 3) = \frac{d}{dt}u(t - 3/2) = \delta(t - 3/2).$$

Since $\frac{d}{dt}(2t - 3) = 2$ we have

$$I_1 = \int_0^2 e^{4t}\frac{1}{2}\delta(t - 3/2)dt = \left[\frac{1}{2}e^{4t}\right]_{t=3/2} = \frac{1}{2}e^6.$$

(ii) Let $I_2 = \int_0^2 e^{4t}\delta(3 - 2t)dt$, so that $x = 3 - 2t$, $t = \frac{1}{2}(3 - x)$, and

$dt = -\frac{1}{2}dx$. This time we get

$$I_2 = \int_3^{-1} e^{2(3-x)}\delta(x)(-\frac{1}{2})dx = \frac{1}{2}\int_{-1}^3 e^{2(3-x)}\delta(x)dx = \frac{1}{2}e^6.$$

Using the alternative approach: $u(3 - 2t) = u(3/2 - t)$ so that

$$\frac{d}{dt}u(3 - 2t) = \frac{d}{dt}u(3/2 - t) = -\delta(t - 3/2).$$

Since $\frac{d}{dt}(3 - 2t) = -2$ this gives

$$I_2 = \int_0^2 e^{4t}\left(-\frac{1}{2}\right)[-\delta(t - 3/2)]dt = \frac{1}{2}e^6 \text{ as before.}$$

(iii) Let $I_3 = \int_{-\infty}^{+\infty} \{\cos t + \sin t\}\delta'(t^3 + t^2 + t)dt.$

If $\phi(t) = t^3 + t^2 + t$ then ϕ is monotone increasing with $\phi(0) = 0$ and so

$$\delta(t^3 + t^2 + t) = \frac{1}{3t^2 + 2t + 1}\frac{d}{dt}u(t^3 + t^2 + t)$$

$$= \frac{1}{3t^2 + 2t + 1}\frac{d}{dt}u(t) = \frac{1}{3t^2 + 2t + 1}\delta(t) = \delta(t) .$$

Further we have

$$\delta'(t^3 + t^2 + t) = \frac{1}{3t^2 + 2t + 1}\frac{d}{dt}\delta(t^3 + t^2 + t) = \frac{1}{3t^2 + 2t + 1}\delta'(t)$$

$$= \delta'(t) - \left[-\frac{6t + 2}{(3t^2 + 2t + 1)^2}\right]\delta(t) = \delta'(t) + 2\delta(t) .$$

Hence, $I_3 = \displaystyle\int_{-\infty}^{+\infty} \{\cos t + \sin t\}(\delta'(t) + 2\delta(t))\,dt$

$$= -[-\sin t + \cos t]_{t=0} + 2[\cos t + \sin t]_{t=0} = 1 .$$

3.7.3 Now let $\phi(t) = (t - \alpha)(t - \beta)$ where $\alpha < \beta$. Then ϕ is monotone decreasing for $-\infty < t < (\alpha + \beta)/2$ and $\phi(\alpha) = 0$; also ϕ is monotone increasing for $(\alpha + \beta)/2 < t < +\infty$ and $\phi(\beta) = 0$. Using the results of Sec. 3.7.1 above we get

$$|\phi'(\alpha)| = |2\alpha - (\alpha + \beta)| = \beta - \alpha; \quad |\phi'(\beta)| = |2\beta - (\alpha + \beta)| = \beta - \alpha \tag{3.37}$$

so that

$$\int_{-\infty}^{(\alpha+\beta)/2} f(\tau)\delta\{\phi(\tau)\}\,d\tau = \frac{f(\alpha)}{\beta - \alpha}; \quad \int_{(\alpha+\beta)/2}^{+\infty} f(\tau)\delta\{\phi(\tau)\}\,d\tau = \frac{f(\beta)}{\beta - \alpha} \tag{3.38}$$

Hence,

$$\int_{-\infty}^{+\infty} f(\tau)\delta\{(t - \alpha)(t - \beta)\}\,d\tau = \frac{1}{\beta - \alpha}\{f(\alpha) + f(\beta)\} . \tag{3.39}$$

To obtain this result from the alternative heuristic approach we first note that

$$\phi(t) > 0 \text{ if } t < \alpha \text{ and if } t > \beta, \text{ while } \phi(t) < 0 \text{ if } \alpha < t < \beta .$$

Hence,

$$u\{(t - \alpha)(t - \beta)\} = u(\alpha - t) + u(t - \beta)$$

and so,

$$\frac{d}{dt}u\{\phi(t)\} = -\delta(\alpha - t) + \delta(t - \beta) = -\delta(t - \alpha) + \delta(t - \beta) .$$

Since $\phi'(t) = 2t - (\alpha + \beta)$ it follows that

$$\delta\{\phi(t)\} = \frac{d}{dt} u\{\phi(t)\} \Big/ \frac{d\phi}{dt} = \frac{-\delta(t-\alpha)}{2t - (\alpha + \beta)} + \frac{\delta(t-\beta)}{2t - (\alpha + \beta)}$$

that is,

$$\delta\{(t-\alpha)(t-\beta)\} = \frac{1}{\beta - \alpha} \{\delta(t-\alpha) + \delta(t-\beta)\}, \quad \alpha < \beta$$

(3.40)

which is equivalent to the result obtained in (3.39).

In particular, if $\beta > 0$ and $\alpha = -\beta$, then

$$\delta(t^2 - \beta^2) = \frac{1}{2\beta} \{\delta(t + \beta) + \delta(t - \beta)\}$$

(3.41)

3.7.4 Let $\phi(t) = \sin t$. Then ϕ is monotone increasing for

$$\left(2m - \frac{1}{2}\right)\pi < t < \left(2m + \frac{1}{2}\right)\pi$$

and $\phi(2m\pi) = 0$; similarly ϕ is monotone decreasing for

$$\left(2m + \frac{1}{2}\right)\pi < t < \left(2m + \frac{3}{2}\right)\pi$$

and $\phi\{(2m + 1)\pi\} = 0$, $(m = 0, \pm1, \pm2, \ldots)$. As a result we can write

$$\int_{\left(2m - \frac{1}{2}\right)\pi}^{\left(2m + \frac{1}{2}\right)\pi} f(\tau)\delta(\sin \tau)d\tau = \frac{1}{|\cos 2m\pi|} f(2m\pi) = f(2m\pi)$$

and

$$\int_{\left(2m + \frac{1}{2}\right)\pi}^{\left(2m + \frac{3}{2}\right)\pi} f(\tau)\delta(\sin \tau)d\tau = \frac{1}{|\cos (2m + 1)\pi|} f\{(2m + 1)\pi\} = f\{(2m + 1)\pi\}.$$

Combining these expressions for all possible values of m gives

$$\int_{-\infty}^{+\infty} f(\tau)\delta(\sin \tau)d\tau = \sum_{n=-\infty}^{+\infty} f(n\pi)$$

(3.42)

provided that the series on the right-hand side converges. The expression $\delta(\sin t)$ therefore *represents an infinite (periodic) train of delta functions* located at the points $t = n\pi$ where $n = 0, \pm1, \pm2, \ldots$. Note that if f is any continuous function which vanishes identically outside some finite interval $[a,b]$ then no problem of convergence arises in (3.42). For in that case there must exist integers n_1, n_2, such that

$$(n_1 - 1)\pi \leqslant a < b \leqslant (n_2 + 1)\pi$$

and at worst we have only to deal with a finite sum

$$\int_{-\infty}^{+\infty} f(\tau)\delta(\sin \tau)\mathrm{d}\tau = \sum_{m=n_1}^{n_2} f(m\pi). \tag{3.43}$$

The alternative heuristic approach is particularly rewarding in this case since, as Fig. 3.2 indicates, it can virtually be carried out by inspection. The detailed steps of the argument are as follows:

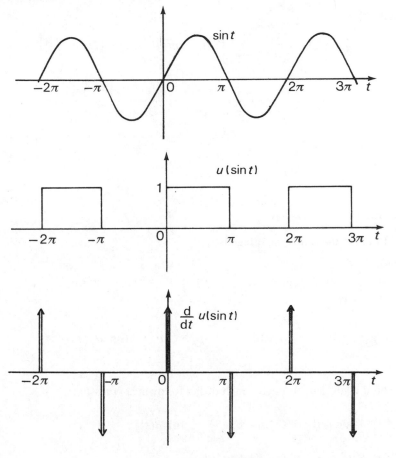

Fig. 3.2.

Since $\phi(t) > 0$ if $2m\pi < t < (2m+1)\pi$ and $\phi(t) < 0$ if $(2m-1)\pi < t < 2m\pi$ for $m = 0, \pm1, \pm2, \ldots$, it follows that

$$u(\sin t) = \sum_{m=-\infty}^{+\infty} \left[u\{t - 2m\pi\} - u\{t - (2m+1)\pi\} \right] .$$

Hence,

$$\frac{d}{dt} u(\sin t) = \sum_{m=-\infty}^{+\infty} \left[\delta\{t - 2m\pi\} - \delta\{t - (2m+1)\pi\} \right]$$

Now $\phi'(t) = \cos t$ and so, for $m = 0, \pm1, \pm2, \ldots$, we have

$$\cos 2m\pi = +1 \quad \text{and} \quad \cos(2m+1)\pi = -1 .$$

Thus,

$$\delta\{\sin t\} = \frac{1}{\cos t} \sum_{m=-\infty}^{+\infty} \left[\delta\{t - 2m\pi\} - \delta\{t - (2m+1)\pi\} \right]$$

$$= \sum_{m=-\infty}^{+\infty} \left[\frac{\delta\{t - 2m\pi\}}{\cos 2m\pi} - \frac{\delta\{t - (2m+1)\pi\}}{\cos(2m+1)\pi} \right]$$

$$= \sum_{m=-\infty}^{+\infty} \delta(t - m\pi) . \tag{3.44}$$

Exercises IV

1. Evaluate the following integrals:

(a) $\displaystyle\int_{-1}^{0} \sinh 2t \, \delta(5t + 2) dt$;   (b) $\displaystyle\int_{-2\pi}^{+2\pi} e^{\pi t} \delta(t^2 - \pi^2) dt$;

(c) $\displaystyle\int_{-\pi}^{\pi} \cosh\theta \, \delta(\cos\theta) d\theta$;   (d) $\displaystyle\int_{-\infty}^{+\infty} e^{-|t|} \delta(\sin \pi t) dt$.

2. Find equivalent forms for each of the following expressions:

(a) $\delta\{\sin |t|\}$;   (b) $\delta\{\cos \dfrac{\pi t}{2}\}$;   (c) $\delta(e^t)$;

(d) $\delta'(\theta^2 - \pi^2)$;   (e) $\delta'\{\sinh 2x\}$.

3.   If x is confined to the range $0 < x < +\infty$, find interpretations for the expressions:

(a) $\delta(\log x)$;   (b) $\delta\left(\dfrac{1}{a} - \dfrac{1}{x}\right)$;

where a is some fixed positive number.

4.* Let ϕ be a twice continuously differentiable function which increases monotonely from $\phi(a)$ to $\phi(b)$ as t increases from a to b. If $\phi(c) = 0$ where $a < c < b$ and $\phi(a) < \phi(c) < \phi(b)$, show that

$$\delta'\{\phi(t)\} = \frac{1}{|\phi'(c)|^2} \left\{\delta'_c(t) + \frac{\phi''(c)}{\phi'(c)} \delta_c(t)\right\}.$$

CHAPTER 4

Time-invariant Linear Systems and Generalised Functions

4.1 SYSTEMS AND OPERATORS

4.1.1 In this chapter we examine in some detail what is perhaps one of the most natural and obvious applications of the generalised functions introduced in Chapters 2 and 3. We begin by stating first what is to be understood by the term "system".

Now a *function* could be defined as a rule, or formula, which maps or transforms *numbers* into *numbers*. An *operator* may be defined similarly as a rule or formula which maps or transforms *functions* into *functions*. By a *system* we shall mean a mathematical model of a physical device which may be represented by (or identified with) an operator, T, which maps the members of one class of functions ("input signals") into those of another ("output signals").

Generally we denote input signals by x_1, x_2, etc., and output signals by y_1, y_2, etc. The relation between a given input signal x and the corresponding output signal y may be indicated either by specific use of an operator symbol T

$$y = T[x] \tag{4.1}$$

or simply by writing

$$x(t) \rightarrow y(t) . \tag{4.2}$$

We shall refer to y as the **image of x under** T or as the **response of the system** represented by the operator T to the input x. Where convenient we shall use the ordinary block diagram representation as shown in Fig. 4.1.

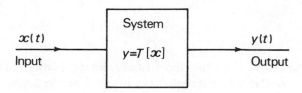

Fig. 4.1.

4.1.2 An operator T is said to be **linear** whenever it enjoys the following two properties:

L1. *Additivity*: If $y_1 = T[x_1]$ and $y_2 = T[x_2]$, then

$$T[x_1 + x_2] = y_1 + y_2 \equiv T[x_1] + T[x_2] \qquad (4.3)$$

L2. *Homogeneity*: If $y = T[x]$ and if α is any real number, then

$$T[\alpha x] = \alpha y \equiv \alpha\{T[x]\} \qquad (4.4)$$

If we say that a system is linear if and only if the operator T which represents it is a linear operator in the sense defined above, then this is the same as saying that a linear system is one for which the Principle of Superposition is satisfied, that is, one for which the response to any finite linear combination of inputs is always the like linear combination of outputs. Thus, if $x_1(t) \to y_1(t)$ and $x_2(t) \to y_2(t)$, and if α and β are any real numbers, then for a linear system we have

$$\alpha x_1(t) + \beta x_2(t) \to \alpha y_1(t) + \beta y_2(t) .$$

Note that from this definition of linearity it follows at once that if $x(t) = 0$ for all t then so also does $y(t)$, that is, there can be no output without some (non-zero) input. This means that there is no stored energy; the definition restricts us to *relaxed* systems. Some examples of linear systems which are of particular importance are given below:

Examples

(i) *Constant gain amplifier*   The output y is always a constant multiple of the input x.

$$x(t) \to y(t) = Ax(t) .$$

(ii) *Multiplier*   The output y is the product of the input x and a certain fixed function ϕ.

$$x(t) \to y(t) = \phi(t)x(t) .$$

(iii) *Differentiator*   The output is the first derivative of the input.

$$x(t) \to y(t) = x'(t) .$$

(iv) *Integrator*   There are several ways in which we might define the ideal integrator as a system in which the input x turns out to be the derivative of the output, y. We shall adopt the following definition:

$$x(t) \to y(t) = \int_{-\infty}^{t} x(\tau)\mathrm{d}\tau.$$

(v) *Ideal Time Delay*   If f is any function of t and if a is any fixed real number, recall that the **translate of f with respect to** a is the function f_a given by

$$f_a(t) = f(t - a) \quad \text{for all } t.$$

If a is positive then the **ideal time delay of duration** a may be defined as the system whose response to any input x is the translate x_a:

$$x(t) \to x(t-a) \equiv x_a(t) .$$

4.1.3 For a model of an actual physical system (however idealised) we would naturally require that all input and output signals should be *real-valued* functions of time. However, it is often convenient to work in terms of *complex-valued signals*. The concept of linearity for a real system can be extended to allow for this eventuality in the following, fairly obvious, way:

L3. *Realness*: Let the responses of the system to the real-valued inputs $x_1(t)$ and $x_2(t)$ be $y_1(t)$ and $y_2(t)$ respectively. Then the response to the complex-valued input $z(t) = x_1(t) + ix_2(t)$ is (defined to be) the complex-valued function $w(t) = y_1(t) + iy_2(t)$.

4.1.4 An operator T is said to be **stationary** if it has the following property.
   If $y = T[x]$, and if a is any fixed real number, then

$$T[x_a] = y_a \equiv \{T[x]\}_a \tag{4.5}$$

Any system represented by a stationary operator T is said to be **time-invariant**. Thus, for a time-invariant system, the only effect of delaying (or advancing) an input signal is to produce a corresponding delay (or advance) of the output signal; if $x(t) \to y(t)$ then,

$$x(t-a) \to y(t-a), \quad \text{for any given } a.$$

   A system which is both linear and time-invariant will often be referred to, for brevity, as a T.I.L.S. (time-invariant linear system). With the exception of the multiplier all the systems described above are both linear and time-invariant. In general, a multiplier is a linear, **time-varying** system; in contrast, the system defined by the input/output relation

$$x(t) \to [x(t)]^2$$

is clearly time-invariant but, equally clearly, non-linear.

4.1.5 Remarks

(i) In practice, many systems of interest are specified by input/output relations which take the form of differential equations or of integro-differential equations. The significance of linearity (and, to some extent, of time-invariance) needs to be examined rather more closely in such cases, because of the problems associated with the specification of initial conditions. We consider this point in detail in the example discussed at the end of this chapter.

(ii) All signals are, at the moment, understood to be *functions* in the usual sense of the word. The philosophy adopted here is that in principle at least it should always be possible to measure the amplitude of any physical signal at any given instant. Hence the use of generalised functions as inputs or outputs is essentially an idealisation and should come logically as an extension of the basic theory. In particular this means that a system will be associated with a specific class of functions which are "admissible" inputs in the sense that the corresponding output functions are well-defined. Thus the differentiator would have as a class of admissible inputs those functions which are differentiable (in the classical sense) everywhere.

(iii) In general a system can be described as "continuous" if, whenever a sequence of input signals converges to zero (in some sense to be specified), the corresponding sequence of output signals also converges to zero (again in some specific sense). In certain circumstances this is equivalent to the demand that "small" inputs should always give rise to correspondingly "small" outputs. In the following heuristic account of linear system theory we will not pursue this question, although we shall indicate from time to time in the text those points where the assumption of system continuity becomes important. For the moment we merely note the following fact.

Linearity is often taken to mean simply that the Principle of Superposition is valid in the sense that the image of the pointwise sum of two admissible inputs is the pointwise sum of the individual responses. Strictly, this only says that the system is additive (L1). However, whenever an appropriate continuity condition is imposed, every additive system turns out to be automatically linear in the full sense of the word.

Exercises I

1. Establish which of the systems defined below in terms of input/output relations is (i) linear, and (ii) time-invariant.

    (a) $x(t) \rightarrow x''(t) - 2x(t)$;

    (b) $x(t) \rightarrow \int_0^t x(\tau)d\tau + x(t)\sin t$;

    (c) $x(t) \rightarrow kx(0)$, where k is a real, non-zero, constant ;

    (d) $x(t) \rightarrow |x(t)|$;

    (e) $x(t) \rightarrow x^+(t)$, where x^+ denotes the *positive* part of x defined by
    $$x^+(t) \equiv \max\{x(t), 0\}.$$

2. If T is an additive operator, prove that:

    (i) for every (admissible) function x we have
    $$T[-x] = -T[x] ,$$

(ii) T is always homogenous with respect to *rational* multipliers; that is to say,

$$T[rx] = rT[x]$$

for every function x and every rational number r.

3.* Let $h(t,\tau)$ be a bounded, continuous function of τ for each fixed t. If x and y are related by the equation

$$y(t) = \int_{-\infty}^{+\infty} x(\tau)h(t,\tau)\,\mathrm{d}\tau$$

show that the mapping $x(t) \to y(t)$ defines a system which is linear (but not necessarily time-invariant). If the system does happen to be time-invariant, show that the function $h(t,\tau)$ actually takes the specific form,

$$h(t,\tau) \equiv h(t-\tau) .$$

4.2 STEP RESPONSE AND IMPULSE RESPONSE

4.2.1 Given a T.I.L.S. represented by an operator T suppose that the unit step function, u, is an admissible input. That is to say in response to the unit step u applied as an input there exists a well-defined function $\sigma = T[u]$ (see Fig. 4.2).

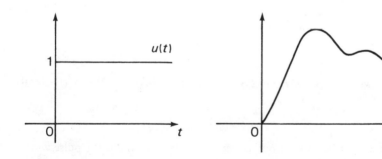

Fig. 4.2(a).   Fig. 4.2(b).

This function σ is usually referred to as the **step-response** of the system.

Now consider the function

$$x(t) = k\,[u(t+a/2) - u(t-a/2)] \qquad (4.6)$$

This represents a rectangular pulse of duration a and of amplitude k, centred about the origin (Fig. 4.3(a)). Since the system concerned is both linear and time-invariant the response to this signal applied as an input can be expressed as a linear combination of suitably translated versions of the step-response (see Fig. 4.3(b)):

$$y(t) \equiv T[x](t) = T[k\{u(t + a/2) - u(t - a/2)\}]$$

$$= k\{T[u(t + a/2)] - T[u(t - a/2)]\}$$

$$= k\{\sigma(t + a/2) - \sigma(t - a/2)\} . \qquad (4.7)$$

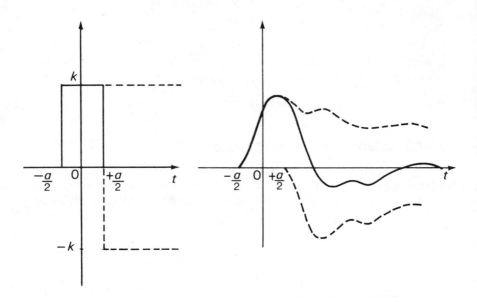

$$x(t) = k\{u(t + a/2) - u(t - a/2)\} \qquad y(t) = k\{\sigma(t + a/2) - \sigma(t - a/2)\}$$

Fig. 4.3(a).   Fig. 4.3(b).

If the function σ has a continuous derivative, $h = \sigma'$, and if the pulse width a is sufficiently small, then we can write

$$k[\sigma(t + a/2) - \sigma(t - a/2)] \simeq (ka)h(t) \qquad (4.8)$$

(For, if σ is continuously differentiable, then the first mean value theorem of the differential calculus gives

$$\sigma(t + a/2) - \sigma(t - a/2) = \sigma'(t + \xi)$$

where ξ is some point such that $-a/2 < \xi < +a/2$. If a is small then by the continuity of σ' we have

$$\sigma'(t + \xi) = \sigma'(t) + \epsilon \simeq \sigma'(t) .)$$

Taking the particular case when $k = 1/a$ we can say that the function h, defined here as the derivative of the step response σ, represents to a first approximation the system response to a narrow pulse of unit area located at the origin. This function h is called the **impulse response** of the system.

4.2.2 Now take for the input x a bounded, continuous, function which vanishes identically outside some finite time-interval, say $\alpha \leqslant t \leqslant \beta$. We can approximate $x(t)$ by a train of adjacent narrow pulses as illustrated in Fig. 4.4.

$$x(t) \simeq \sum_k x(\tau_k)[u(t - t_k) - u(t - t_{k-1})].$$

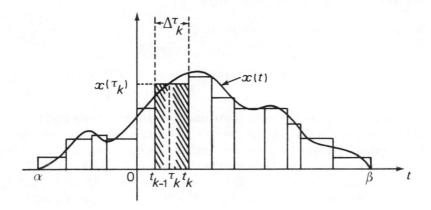

Fig. 4.4.

The response of the system to an elementary pulse of width $\Delta\tau_k$, where $\Delta\tau_k = t_k - t_{k-1}$, and of height $x(\tau_k)$, centred about the point $t = \tau_k$, will be given (approximately) by

$$\{x(\tau_k)\Delta\tau_k\}h(t - \tau_k),$$

using (4.8), and the fact that the system is time-invariant. Next, by linearity of the system, the response to the input $x(t)$ will be given, to a first approximation, by the sum

$$\sum_k x(\tau_k)h(t - \tau_k)\Delta\tau_k. \tag{4.9}$$

In the limit, as the approximating pulses are made narrower and narrower, we obtain the actual output y in the form of a so-called **convolution integral**:

$$y(t) = \int_\alpha^\beta x(\tau)h(t - \tau)\,\mathrm{d}\tau = \int_{-\infty}^{+\infty} x(\tau)h(t - \tau)\,\mathrm{d}\tau. \tag{4.10}$$

We can take infinite limits in (4.10) just because the function x is known to vanish identically outside the finite interval $[\alpha, \beta]$. If we remove this constraint on x then an extension of the above argument can be used to show that the output y will still be given by an integral of the form (4.10), provided that the functions x and h are suitably well-behaved for large values of $|t|$ (so that the infinite integral does converge). Similarly, we may weaken considerably the assumption that x and h are continuous. Note, however, that the one crucial step in the argument is the passage to the limit in the derivation of (4.10) from (4.9). It is this step which relies explicitly on a suitable assumption of system continuity as discussed in the Remarks at the end of section (4.1) above.

4.3 CONVOLUTION AS AN OPERATION

4.3.1 By a simple change of variable we have

$$y(t) = \int_{-\infty}^{+\infty} x(\tau)h(t-\tau)\,d\tau = \int_{-\infty}^{+\infty} x(t-\tau)h(\tau)\,d\tau . \qquad (4.11)$$

This *commutative* property of the convolution integral effectively says that the input function x and the impulse response function h are, in a certain sense, interchangeable. We could equally well regard y as the response of a system, whose impulse response function is x, to the input signal h.

Now consider two T.I.L.S., with impulse response functions h_1 and h_2 respectively, connected in cascade (Fig. 4.5). The response z to a continuous input x can be computed as follows (always assuming that the change in the order of integration can be justified).

$$z(t) = \int_{-\infty}^{+\infty} y(t-\theta)h_2(\theta)\,d\theta = \left[\int_{-\infty}^{+\infty} h_2(\theta) \int_{-\infty}^{+\infty} x(t-\theta-\phi)h_1(\phi)\,d\phi\right]d\theta$$

$$= \int_{-\infty}^{+\infty} h_2(\theta)\left[\int_{-\infty}^{+\infty} x(t-\tau)h_1(\tau-\theta)\,d\tau\right]d\theta , \quad \text{putting } \tau = \theta + \phi ,$$

$$= \int_{-\infty}^{+\infty} x(t-\tau)h_3(\tau)\,d\tau , \quad \text{where } h_3(\tau) = \int_{-\infty}^{+\infty} h_1(\tau-\theta)h_2(\theta)\,d\theta .$$

Thus the cascade connection of the two T.I.L.S. is equivalent to a single T.I.L.S. whose impulse response is the convolution of the individual impulse response functions of the connected systems.

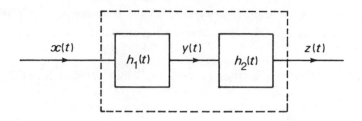

Fig. 4.5.

4.3.2 As an operation on functions, it is usual to denote convolving by the symbol *:

$$h_3(t) = h_1(t) * h_2(t) \equiv \int_{-\infty}^{+\infty} h_1(t - \tau) h_2(\tau) d\tau . \qquad (4.12)$$

The above results show that, whenever the appropriate convolutions are defined, the operation is both commutative and associative:

$$(h_1 * h_2)(t) = (h_2 * h_1)(t)$$

$$\{h_1 * (h_2 * h_3)\}(t) = \{(h_1 * h_2) * h_3\}(t) .$$

Likewise it is easy to confirm that the distributive law holds:

$$\{(h_1 + h_2) * h_3\}(t) = (h_1 * h_3)(t) + (h_2 * h_3)(t) .$$

Thus, from a purely formal point of view, the operation of convolving functions has much in common with multiplication.

Exercises II

1. Find the step response and the impulse response of the ideal integrator, defined by the input/output relation $y(t) = \int_{-\infty}^{t} x(\tau) d\tau$.

2. Evaluate each of the following convolutions:

    (a) $u(t) * u(t)$;   (b) $e^{\alpha t} * \{u(t) \cos \omega t\}$;

    (c) $\{u(t) \sin t\} * \{u(t) \cos t\}$;   (d) $p(t) * p(t)$;

    where the function p is given by $p(t) = u\left(t + \frac{1}{2}\right) - u\left(t - \frac{1}{2}\right)$.

3. Let f be a continuous function which vanishes for all $t < 0$; define g_n as follows:

$$g_n(t) = \int_0^t \frac{(t - \tau)^n}{n!} f(\tau) d\tau , \quad \text{for } n \geqslant 1 ; \quad g_0(t) = \int_0^t f(\tau) d\tau .$$

Show that g_n is the result of integrating f $(n + 1)$ times from 0 to t. [Hint: use the Leibniz formula for differentiating an integral with respect to a parameter.]

4. Let f be an arbitrary continuous function which vanishes for all $t < 0$; if $g(t) = u(t) \dfrac{1}{\sqrt{\pi t}}$, prove that $f(t)*g(t)$ is the "half-integral" of f from 0 to t, in the sense that

$$\{f(t)*g(t)\}*g(t) = \int_0^t f(\tau)\,d\tau .$$

[Hint: first evaluate $g(t)*g(t)$.]

4.4 GENERALISED IMPULSE RESPONSE FUNCTIONS

4.4.1 Consider in particular the case of the time-invariant linear system which has the property of leaving every input signal x unchanged. If we assume that there exists a convolution integral representation for this system, then the analogue of equation (4.10) should take the form

$$x(t) = \int_{-\infty}^{+\infty} x(t-\tau)h(\tau)\,d\tau = \int_{-\infty}^{+\infty} x(\tau)h(t-\tau)\,d\tau \qquad (4.13)$$

where h denotes the impulse response function characterising the system, and x is an arbitrary continuous function. Putting $t = 0$ in (4.13) we get

$$x(0) = \int_{-\infty}^{+\infty} x(-\tau)h(\tau)\,d\tau = \int_{-\infty}^{+\infty} x(\tau)h(-\tau)\,d\tau . \qquad (4.14)$$

Now write $z(t) \equiv x(-t)$, and apply (4.14) directly to the function z in place of x:

$$z(0) = \int_{-\infty}^{+\infty} z(-\tau)h(\tau)\,d\tau \equiv \int_{-\infty}^{+\infty} x(\tau)h(\tau)\,d\tau . \qquad (4.15)$$

But, since $z(0) = x(0)$, this last equation is simply a restatement of the fundamental sampling property of the delta function, and it follows that the system in question must have the delta function itself as its impulse response function. (This is, of course, consistent with the fact that the unit step function, applied as an input, would be transmitted unchanged; that is, the step response of the system is the unit step function, u.) More generally the impulse response function associated with an ideal time delay of duration a is seen to be the shifted delta function, δ_a:

$$x(t - a) = \int_{-\infty}^{+\infty} x(t - \tau)\delta_a(\tau)\,d\tau = f(t)*\delta_a(t) . \tag{4.16}$$

4.4.2 A similar argument applies to the ideal differentiator. If a continuously differentiable input function x is applied to this system, then a formal appeal to equation (4.10) yields the result

$$x'(t) = \int_{-\infty}^{+\infty} x(t - \tau)h(\tau)\,d\tau = \int_{-\infty}^{+\infty} x(\tau)h(t - \tau)\,d\tau . \tag{4.17}$$

Thus, putting $t = 0$, we have

$$x'(0) = \int_{-\infty}^{+\infty} x(-\tau)h(\tau)\,d\tau = \int_{-\infty}^{+\infty} x(\tau)h(-\tau)\,d\tau . \tag{4.18}$$

Writing $z(t) \equiv x(-t)$, as before, and applying (4.18) to the function z,

$$z'(0) = \int_{-\infty}^{+\infty} z(-\tau)h(\tau)\,d\tau \equiv \int_{-\infty}^{+\infty} x(\tau)h(\tau)\,d\tau . \tag{4.19}$$

This time, however, we note that

$$\frac{d}{dt} z(t) = \frac{d}{dt} x(-t) = -\frac{d}{d(-t)} x(-t)$$

so that, in particular, $z'(0) = -x'(0)$, and (4.19) is now seen to be the characteristic sampling property associated with the first derivative, δ', of the delta function.

In the same way, the generalised function $\delta^{(n)}$ may be interpreted as the impulse response function characterising the time-invariant linear system which transforms every n-times continuously differentiable input function x into its nth derivative $x^{(n)}$.

4.4.3 These results lead naturally to the extension of the operation of convolution to include generalised functions. In Sec. 4.3 we have seen that the convolution of two ordinary functions, $h_1 * h_2$, can be interpreted in terms of the cascade connection of the time-invariant linear systems of which h_1 and h_2 are the impulse response functions. If we replace one of these functions by a delta function, then the convolution $h_1 * \delta_a$ (where h_1 is assumed to be continuous) can be interpreted as either

(i) the impulse response function corresponding to the cascade connection of the system with impulse response function h_1 and of a time delay of duration a, or

(ii) the response of a time delay of duration a to an input signal h_1, or

(iii) the response of a system with impulse response function h_1 to the delta function δ_a applied as an input.

The last of these three alternatives is the one most often encountered in the classical literature of linear systems theory, and is, indeed, the genesis of the name "impulse response" for the function h which we have defined here as the derivative of the step response.

As for the convolution of two delta functions, say $\delta_a * \delta_b$, note that the cascade connection of two time delays of durations a and b respectively ought to result in a single time delay of duration $(a + b)$. That is to say, we should have

$$\delta_a(t) * \delta_b(t) = \int_{-\infty}^{+\infty} \delta_a(t-\tau)\delta_b(\tau)d\tau = \delta_a(t-b) = \delta_{a+b}(t) . \quad (4.20)$$

That this is the case can be seen from the following formal manipulations:

$$\int_{-\infty}^{+\infty} \delta_b(\theta)\cdot\left[\int_{-\infty}^{+\infty} f(t-\theta-\phi)\delta_a(\phi)d\phi\right]d\theta$$

$$= \int_{-\infty}^{+\infty} \delta_b(\theta)f(t-\theta-a)d\theta = f(t-b-a) .$$

With regard to the derivatives of the delta function, let f be a function which has continuous derivatives up to at least the $(m + n)^{\text{th}}$ order. Then we have

$$[f(t) * \delta^{(n)}(t)] * \delta^{(m)}(t) = f^{(n)}(t) * \delta^{(m)}(t) = f^{(n+m)}(t)$$

and it follows that

$$\delta^{(n)}(t) * \delta^{(m)}(t) = \int_{-\infty}^{+\infty} \delta^{(n)}(t-\tau)\delta^{(m)}(\tau)d\tau = \delta^{(m+n)}(t) \quad (4.21)$$

for any $m \geqslant 0$ and any $n \geqslant 0$.

4.5 TRANSFER FUNCTION AND FREQUENCY RESPONSE

4.5.1 Let $H_s(t)$ denote the response of a given T.I.L.S. to the input signal $x(t) = e^{st}$:

$$e^{st} \to H_s(t) .$$

Since the system is **time-invariant** we know that, for any given fixed value of τ, the response to the input $x(t + \tau)$ must be $H_s(t + \tau)$. But if $x(t) = e^{st}$ then

$$x(t + \tau) = e^{s(t+\tau)} = e^{s\tau}e^{st} = e^{s\tau}x(t) \ .$$

Hence, since the system is *linear*, the response to the input $x(t + \tau)$ must be $e^{s\tau}H_s(t)$ and so we must have

$$H_s(t + \tau) = e^{s\tau}H_s(t) \ .$$

In particular we can put $t = 0$ to get

$$H_s(\tau) = e^{s\tau}H_s(0) \ . \tag{4.22}$$

The relation (4.22) holds for any given value of τ. Writing t (a variable) in place of τ, and denoting the number $H_s(0)$ by the symbol $H(s)$, (since it depends only on the value of the parameter s) we get the following result:

The response of a (relaxed) T.I.L.S. to an exponential input, $x(t) = e^{st}$, is always of the form

$$y(t) = H(s)e^{st} \tag{4.23}$$

where $H(s)$ is a number which depends only on the particular system concerned and on the parameter s.

Allowing the parameter s in (4.23) to take complex values, $s = \sigma + i\omega$, we refer to $H(s)$ as the **transfer function** of the system concerned. Note that for a given system there will generally be inputs of the form e^{st} for which there exists no well-defined response; equivalently, the transfer function $H(s)$ will be defined in general only for a limited range of values of s (that is, only on some region of the complex plane).

4.5.2 Now suppose that the T.I.L.S. is one for which there exists a well-defined impulse response function $h(t)$. Using the convolution integral (4.10), the response of the system to the exponential input e^{st} is given by

$$\int_{-\infty}^{+\infty} e^{s\tau}h(t - \tau)\mathrm{d}\tau = \int_{-\infty}^{+\infty} e^{s(t-\tau)}h(\tau)\mathrm{d}\tau$$

$$= e^{st}\int_{-\infty}^{+\infty} e^{-s\tau}h(\tau)\mathrm{d}\tau \tag{4.24}$$

Comparing (4.23) and (4.24) we obtain the following explicit representation for the transfer function, $H(s)$,

$$H(s) = \int_{-\infty}^{+\infty} e^{-s\tau} h(\tau) d\tau \qquad (4.25)$$

whenever the infinite integral exists.

Equation (4.25) defines what is usually described as the **two-sided Laplace Transform** of the function h. If we put $s = i\omega$ then, again provided the infinite integral exists, we obtain the so-called **Fourier Transform** of the function h:

$$H(i\omega) = \int_{-\infty}^{+\infty} e^{-i\omega\tau} h(\tau) d\tau . \qquad (4.26)$$

In most modern texts it is usual to define the **one-sided Laplace Transform** of the function h as the following:

$$\mathcal{L}\{h(t)\} \equiv \int_{0}^{\infty} e^{-s\tau} h(\tau) d\tau . \qquad (4.27)$$

Note that this may be interpreted as the two-sided transform of the function $u(t)h(t)$:

$$\mathcal{L}\{h(t)\} = \int_{0}^{\infty} e^{-s\tau} h(\tau) d\tau = \int_{-\infty}^{+\infty} e^{-s\tau} \{u(\tau)h(\tau)\} d\tau . \qquad (4.28)$$

In the case of a system whose impulse response function h vanishes for all negative values of t we have

$$y(t) = \int_{-\infty}^{t} x(\tau)h(t-\tau) d\tau = \int_{0}^{\infty} x(t-\tau)h(\tau) d\tau \qquad (4.29)$$

and it follows at once that if the input x vanishes for all $t < t_0$ then so also does the output y. A system of this kind is said to be **causal**, or **non-anticipative**, in the sense that no output signal can ever precede the input signal which gives rise to it. Accordingly the one-sided Laplace Transform of an arbitrary function h could always be interpreted as the transfer function of the *causal* system whose impulse response function is $u(t)h(t)$. Note also that, for a causal system, if the input signal x is such that $x(t) = 0$ for all $t < 0$ then the output signal y is given by the following expression:

$$y(t) = \int_{0}^{t} x(\tau)h(t-\tau) d\tau = \int_{0}^{t} x(t-\tau)h(\tau) d\tau . \qquad (4.30)$$

(In some texts (4.30) is actually used as the definition of the convolution of two arbitrary functions x and h, whether those functions vanish for negative values of t or not. We shall not adopt this convention here.)

4.5.3 The remarks in Sec. 4.4 on the appearance of generalised functions as impulse responses suggests that Laplace and Fourier Transforms for delta functions and derivatives of delta functions could be defined in terms of the appropriate transfer functions. So far as the two-sided Laplace Transform and the Fourier Transform are concerned there is, of course, no particular merit in this since we may appeal directly to the characteristic sampling properties in each case:

$$\int_{-\infty}^{+\infty} e^{-st}\delta(t)\,dt = \left[e^{-st}\right]_{t=0} = 1$$

$$\int_{-\infty}^{+\infty} e^{-st}\delta'(t)\,dt = -\left[\frac{d}{dt}(e^{-st})\right]_{t=0} = s$$

and so on. In the case of the one-sided Laplace Transform, however, we are confronted with expressions of the form

$$\int_0^\infty e^{-st}\delta(t)\,dt \quad \text{and} \quad \int_0^\infty e^{-st}\delta'(t)\,dt$$

which are, strictly, undefined in the formal calculus of delta functions. If we adopt the point of view suggested above, then the Laplace Transform of the delta function would be identified with the transfer function of the *causal* T.I.L.S. which transmits every input signal unchanged. In this sense we would be entitled to write

$$\mathcal{L}\{\delta(t)\} = 1 .$$

Similarly, treating δ' as the impulse response of the ideal differentiator and noting that this is most certainly a causal system, we should have

$$\mathcal{L}\{\delta'(t)\} = s .$$

Exercises III

1.   Evaluate the integral

$$\int_{-\infty}^{+\infty} \{x^4 - 10x^2 + 1\}\{\delta_{\sqrt{2}}(x)*\delta_{\sqrt{3}}(x)\}\,dx .$$

2. Find the frequency response functions associated with the following time-invariant linear systems:

    (a) an ideal amplifier of constant gain,
    (b) an ideal time delay,
    (c) the differentiator,
    (d) the integrator (as defined in Ex. II, No. 1).

3. A linear, time-invariant system has step-response function $\sigma(t) = u(t) \sin \alpha t$. Find the output of the system if the input is the function $x(t) = u(t) \sin \alpha t$. What is the frequency response function of the system?

4.6 EXAMPLE: A SYSTEM REPRESENTED BY A DIFFERENTIAL EQUATION

4.6.1 The electrical circuit shown in Fig. 4.6 can be described in terms of the first-order differential equation

$$v(t) = R \frac{dq}{dt} + \frac{1}{C} q(t) \tag{4.31}$$

where $q(t)$ denotes the charge on the capacitor at time t and $v(t)$ the value of the voltage impressed on the circuit by an independent external generator.

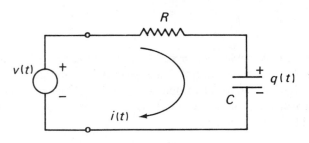

Fig. 4.6.

We can write this equation in the form

$$v = \mathcal{A}q \tag{4.32}$$

where $\mathcal{A} \equiv R \dfrac{d}{dt} + \dfrac{1}{C}$. This symbol $\mathcal{A}$ is an operator in that it represents a transformation which carries the function q into the function v; as is easily confirmed this operator is both linear and stationary. However, this operator formulation of equation (4.31) does not correspond to the way in which the physical situation is understood when we think of it in terms of a *system*. Equation

(4.32) exhibits the voltage v as the result of operating on the charge q. We would more commonly think of the generator as supplying a *known* voltage wave-form v and seek to find the *resulting* charge q. In brief, the system which the electrical circuit of Fig. 4.6 represents is usually thought of as that which is described by the *solution* of the differential equation (4.31).

Now equation (4.31) is a linear first-order equation with integrating factor $\exp(t/CR)$. Its explicit solution presents no particular difficulty, but it is instructive to see precisely what is involved in carrying it out. We have

$$\frac{d}{dt}\{q(t)e^{t/CR}\} \equiv e^{t/CR}\frac{dq}{dt} + \frac{1}{CR}e^{t/CR}q(t) = \frac{v(t)}{R}e^{t/CR}$$

so that

$$q(t)e^{t/CR} - q(t_0)e^{t_0/CR} = \int_{t_0}^{t}\frac{v(\tau)}{R}e^{\tau/CR}\,d\tau \tag{4.33}$$

If we make the very reasonable assumption that the charge q_0 remains bounded as t_0 tends to $-\infty$, then we must have

$$\lim_{t_0 \to -\infty}\{q(t_0)e^{t_0/CR}\} = 0$$

Hence, allowing t_0 to tend to $-\infty$ in (4.33), we get

$$q(t) = \frac{1}{R}\int_{-\infty}^{t}v(\tau)e^{-(t-\tau)/CR}\,d\tau \tag{4.34}$$

which we can write in the form

$$q(t) = \int_{-\infty}^{t}v(\tau)h(t-\tau)\,d\tau \equiv v(t) * h(t) \tag{4.35}$$

where

$$h(t) = \frac{u(t)}{R}e^{-t/CR}.$$

This expresses q explicitly in terms of v, and the relationship can once again be written in operator form:

$$q = \mathcal{C}v \tag{4.36}$$

The operator $\mathcal{C}$ is linear and time-invariant and admits the convolution integral representation (4.34). The function h is the impulse response function and represents the charge on the capacitor due to an impulsive (delta function) voltage excitation at time $t = 0$. Since $h(t)$ vanishes for all $t < 0$ the system is a causal one, as indeed the physical background of the situation would lead us to expect.

4.6.2 In practice we are most often interested in examining the behaviour of the system from some given instant onwards, say for all $t > 0$. Taking $t_0 = 0$ in (4.33) we would get

$$q(t) = \frac{1}{R} \int_0^t v(\tau) e^{-(t-\tau)/CR} \, d\tau + q_0 e^{-t/CR} \qquad (4.37)$$

where we write q_0 for $q(0)$, the charge on the capacitor at time $t = 0$. If v is taken to be a "suddenly applied" excitation voltage, so that $v(t) = 0$ for all $t < 0$, then q_0 is an arbitrarily specified quantity which represents the effect of any energy stored in the system at time $t = 0$. Note that unless this initial charge is zero the relationship between q and v expressed by (4.37) is not linear. Thus, suppose that the input voltage is a unit step, say $v_1(t) = u(t)$. Then for $t > 0$ we get

$$q_1(t) = q_0 e^{-t/CR} + \frac{e^{-t/CR}}{R} \int_0^t e^{\tau/CR} \, d\tau = q_0 e^{-t/CR} + C \left[1 - e^{-t/CR} \right].$$

On the other hand, if $v_2(t) = 2u(t)$ then

$$q_2(t) = q_0 e^{-t/CR} + 2C \left[1 - e^{-t/CR} \right] \neq 2q_1(t) \quad \text{unless } q_0 = 0.$$

The classical Laplace transform is peculiarly well suited to problems in which initial conditions must be taken into account. In the next chapter we shall give a brief review of the classical one-sided Laplace Transform (hereinafter referred to simply as *the* Laplace Transform) together with its natural extension to the generalised functions introduced in what has gone before. In the particular example discussed here note that we can obtain the Laplace Transform of the function h or, equivalently, the Transfer Function of the (causal) system of which h is the impulse response, by computing directly the system response to the input $v(t) = e^{st}$. Thus from (4.34) we get

$$q(t) = \frac{1}{R} \int_{-\infty}^t e^{s\tau} e^{-(t-\tau)/CR} \, d\tau = \frac{C}{1 + sCR} e^{st}$$

so that the transfer function is given by

$$H(s) = \frac{C}{1 + sCR} \equiv \mathcal{L} \left\{ h(t) \right\}.$$

CHAPTER 5

The Laplace Transform

5.1 THE CLASSICAL LAPLACE TRANSFORM

5.1.1 Let f be a function of the real variable t which is defined for all $t \geqslant 0$ and which is either continuous or at least sectionally continuous. The classical Laplace Transform[†] of f is the function $F_0(s)$ defined by the formula

$$F_0(s) \equiv \mathcal{L}\{f(t)\} = \int_0^\infty e^{-st} f(t)\,dt . \qquad (5.1)$$

This definition of $F_0(s)$ clearly makes sense only for those values of s for which the infinite integral is convergent. For many applications it is enough to regard s as a real parameter, but in general it should be taken as complex, say $s = \sigma + i\omega$. Thus $F_0(s)$ is really a function of a complex variable defined over a certain region of the complex plane; the region of definition comprises just those values of s for which the infinite integral exists.

5.1.2 For any specific value of s the infinite integral defining $F_0(s)$ will converge if $|f(t)|$ does not increase "too rapidly" as t goes to infinity. A sufficient condition for the existence of $F_0(s)$ can be stated as follows:

Existence Theorem
If real numbers $M > 0$ and γ exist such that

$$|f(t)| \leqslant M e^{\gamma t} \quad \text{for all } t \geqslant T$$

then $F_0(s)$ exists for all s such that $\mathrm{Re}(s) > \gamma$. (In such a case f is said to be a **function of exponential order** as $t \to \infty$.)

[†] We use the notation $F_0(s)$ for the one-sided Laplace Transform of f and reserve $F(s)$ for the two-sided Laplace Transform. This is consistent with the use of $F(i\omega)$ for the Fourier Transform of f.

Proof

$$F_0(s) = \int_0^\infty e^{-st} f(t)\,dt = \int_0^T e^{-st} f(t)\,dt + \int_T^\infty e^{-st} f(t)\,dt$$

and
$$\left| \int_T^\infty e^{-st} f(t)\,dt \right| \leqslant \int_T^\infty |e^{-st} f(t)|\,dt \leqslant \int_T^\infty e^{-\sigma t} |f(t)|\,dt$$

$$\leqslant \int_T^\infty M e^{-(\sigma - \gamma)t}\,dt \leqslant \int_0^\infty M e^{-(\sigma - \gamma)t}\,dt\ .$$

Provided that $\sigma \equiv \mathrm{Re}(s) > \gamma$, this last integral has the finite value $M/(\sigma - \gamma)$. Hence $F_0(s)$ certainly exists for any s such that $\mathrm{Re}(s) > \gamma$, and the defining integral (5.1) converges absolutely for all such values of s.

Unless otherwise stated we shall assume throughout that every function f which has a well-defined Laplace Transform $F_0(s)$ (over some region of the complex plane) is a function of exponential order as $t \to \infty$. In particular this means that we always have

$$\lim_{t \to \infty} e^{-st} f(t) = 0 \qquad\qquad (5.2)$$

for every value of s at which $F_0(s)$ is defined.

5.1.3 Elementary applications of the Laplace Transform depend essentially on three basic properties:

(i) *Linearity*. If the Laplace Transforms of f and g are $F_0(s)$ and $G_0(s)$ respectively, and if a_1 and a_2 are any (real) constants, then the Laplace Transform of the function h defined by

$$h(t) = a_1 f(t) + a_2 g(t)$$

is
$$H_0(s) = a_1 F_0(s) + a_2 G_0(s)\ .$$

The proof is trivial.

(ii) *Transform of a Derivative*. If f is differentiable (and therefore continuous) for $t \geqslant 0$, then

$$\mathcal{L}\left\{f'(t)\right\} = s F_0(s) - f(0)\ . \qquad\qquad (5.3)$$

Proof

Using integration by parts we have

$$\mathcal{L}\left\{f'(t)\right\} = \int_0^\infty e^{-st} f'(t)\,dt = \left[e^{-st} f(t)\right]_0^\infty + \int_0^\infty s e^{-st} f(t)\,dt$$

$$= -f(0) + s \int_0^\infty e^{-st} f(t)\,dt$$

since $\lim_{t \to \infty} e^{-st} f(t) = 0\ .$

Corollary. If f is n-times differentiable for $t \geqslant 0$, then

$$\mathcal{L}\{f^{(n)}(t)\} = s^n F_0(s) - s^{n-1}f(0) - s^{n-2}f'(0)\ldots\ldots - f^{(n-1)}(0).$$

(iii) *The Convolution Theorem.* Let f and g have Laplace Transforms $F_0(s)$ and $G_0(s)$ respectively, and define h as follows:

$$h(t) = \int_0^t f(\tau)g(t-\tau)d\tau, \quad t \geqslant 0.$$

Then,

$$\mathcal{L}\{h(t)\} = F_0(s)G_0(s). \tag{5.4}$$

(Recall that h, as defined here, is the convolution of the functions $u(t)f(t)$ and $u(t)g(t)$. If f and g happen to be functions which vanish identically for all negative values of t then the above result can be expressed in the form:

The Laplace Transform of the convolution of f and g is the product of the individual Laplace Transforms.)

Proof

The Laplace Transform of h is given by

$$H_0(s) = \int_0^\infty e^{-st}\left[\int_0^t f(\tau)g(t-\tau)d\tau\right]dt.$$

Now,

$$\int_0^t f(\tau)g(t-\tau)d\tau = \int_0^\infty f(\tau)g(t-\tau)u(t-\tau)d\tau$$

because

$$u(t-\tau) = 1 \quad \text{for all } \tau \text{ such that } \tau < t$$

and

$$u(t-\tau) = 0 \quad \text{for all } \tau \text{ such that } \tau > t.$$

Hence

$$H_0(s) = \int_0^\infty e^{-st}\left[\int_0^\infty f(\tau)g(t-\tau)u(t-\tau)d\tau\right]dt.$$

Again,

$$\int_0^\infty g(t-\tau)u(t-\tau)e^{-st}dt = \int_\tau^\infty g(t-\tau)e^{-st}dt$$

because

$$u(t-\tau) = 1 \quad \text{for all } t \text{ such that } t > \tau,$$

and

$$u(t-\tau) = 0 \quad \text{for all } t \text{ such that } t < \tau.$$

Thus,

$$H_0(s) = \int_0^\infty f(\tau)\left[\int_\tau^\infty g(t-\tau)e^{-st}dt\right]d\tau$$

and so, putting $T = t - \tau$, we get

$$H_0(s) = \int_0^\infty f(\tau) \left[\int_0^\infty g(T) e^{-s(T+\tau)} dT \right] d\tau$$

since $T = 0$ when $t = \tau$.

That is,

$$H_0(s) = \int_0^\infty f(\tau) e^{-s\tau} d\tau \int_0^\infty g(T) e^{-sT} dT = F_0(s) G_0(s) .$$

Remark. The change in the order of integration in the proof given above is justified by the absolute convergence of the integrals concerned.

5.1.4 The most immediate application of these properties is in the solution of ordinary differential equations with constant coefficients. Consider the case of the general second-order equation

$$a \frac{d^2 y}{dt^2} + 2b \frac{dy}{dt} + cy = f(t) \tag{5.5}$$

where $y(0) = \alpha$ and $y'(0) = \beta$. If $\mathcal{L}\{y(t)\} = Y_0(s)$ then

$$\mathcal{L}\left\{ \frac{dy}{dt} \right\} = sY_0(s) - \alpha, \quad \text{and} \quad \mathcal{L}\left\{ \frac{d^2 y}{dt^2} \right\} = s^2 Y_0(s) - \alpha s - \beta .$$

Taking Laplace Transforms of both sides of (5.5) therefore gives

$$a\{s^2 Y_0(s) - \alpha s - \beta\} + 2b\{sY_0(s) - \alpha\} + c Y_0(s) = F_0(s) .$$

That is,

$$Y_0(s) = \frac{F_0(s)}{as^2 + 2bs + c} + \frac{a\alpha s + (a\beta + 2b\alpha)}{as^2 + 2bs + c} . \tag{5.6}$$

$Y_0(s)$ is thus given explicitly as a function of s, and what remains is an **inversion problem**; that is to say we need to determine a function $y(t)$ whose Laplace Transform is $Y_0(s)$. The question of uniqueness which naturally arises at this point is not, in practice, a serious problem. In brief, if y_1 and y_2 are any two functions which have the same Laplace transform $Y_0(s)$, then they can differ in value only on a set of points which is (in a sense which can be made precise) a negligibly small set. In fact we have the following situation:

if $\qquad \mathcal{L}\{y_1(t)\} = \mathcal{L}\{y_2(t)\} \quad$ then $\quad \int_0^\infty |y_1(t) - y_2(t)| dt = 0 .$

With this proviso in mind, we admit the slight abuse of notation involved, and write:

$$y(t) \equiv \mathcal{L}^{-1}\{Y_0(s)\} = \mathcal{L}^{-1}\left\{\frac{F_0(s)}{as^2 + 2bs + c}\right\} + \mathcal{L}^{-1}\left\{\frac{a\alpha s + (a\beta + 2b\alpha)}{as^2 + 2bs + c}\right\} \quad (5.7)$$

where y is defined for all $t > 0$.

A more serious problem from the practical point of view is that of implementing the required inversion; that is, of devising effective procedures which allow us to recover a function $f(t)$ given its Laplace Transform $F_0(s)$. In a large number of commonly occurring cases this can be done by expressing $F_0(s)$ as a combination of standard functions of s whose inverse transforms are known (see Sec. 5.4 below).

5.1.5 Note that with zero initial conditions, ($y(0) = y'(0) = 0$), the differential equation (5.5) can be regarded as representing a linear time-invariant system which transforms a given input signal f into a corresponding output y. This output function y is the **particular integral** associated with f and, using the Convolution Theorem, it can be expressed in terms of the appropriate impulse response function characterising the system:

$$y(t) = \int_0^t f(\tau)h_1(t - \tau)\mathrm{d}\tau = \mathcal{L}^{-1}\{F_0(s)H_0(s)\}$$

where

$$H_0(s) = \int_0^\infty e^{-st}h(t)\mathrm{d}t = \frac{1}{as^2 + 2bs + c}.$$

Non-zero initial conditions correspond to the presence of stored energy in the system at time $t = 0$. The response of the system to this stored energy is independent of the particular input f and is given by the **complementary function**. The complete solution (valid for all $t > 0$) of the equation (5.5) can be written in the form

$$y(t) = \mathcal{L}^{-1}\{F_0(s)H_0(s)\} + \mathcal{L}^{-1}\{[a\alpha s + (a\beta + 2b\alpha)] H_0(s)\}. \quad (5.8)$$

In applying the classical Laplace transform technique to (5.5) we are tacitly assuming that the system which it is being taken to represent is **unforced** for $t < 0$; that is, that the response which we compute from (5.8) is actually the response to the excitation $f(t)u(t)$. This is sometimes expressed by saying that the input is **suddenly applied** at time $t = 0$.

Exercises I

1. Find (from first principles, or otherwise) the Laplace Transforms of each of the following functions:

    (a) $u(t)$;   (b) $u(t) - u(t-1)$;   (c) e^{at} ;   (d) t ;
    (e) $\cosh bt$;   (f) $\sin \omega t$;   (g) $\cos \omega t$.

2. Use Laplace Transforms to obtain, for $t \geqslant 0$, the solution of the linear differential equation

$$\frac{d^2 y}{dt^2} + y = t$$

which satisfies the conditions $y(0) = 1$, $y'(0) = -2$.

3. Use the Convolution Theorem for the Laplace Transform to solve the integral equation

$$y(t) = \cos t + 2\sin t + \int_0^t y(\tau)\sin(t-\tau)d\tau$$

for $t > 0$.

4. Solve the simultaneous linear differential equations (for $t > 0$),

$$\frac{dx}{dt} = 2x - 3y$$

$$\frac{dy}{dt} = y - 2x$$

given that $x(0) = 8$, and $y(0) = 3$.

5. A causal, time-invariant, linear system has impulse response function

$$g(t) = \frac{u(t)}{\sqrt{\pi t}}.$$

Find its frequency response function. (Hint: see Exercises II, No. 4, of Chapter 4.)

5.2 LAPLACE TRANSFORMS OF GENERALISED FUNCTIONS

5.2.1 If a is any positive number then there is no especial difficulty in extending the definition of the classical, one-sided, Laplace Transform to apply to the case of a delta function located at $t = a$, or to any of its derivatives located there; for a direct application of the appropriate sampling property gives immediately

$$\mathcal{L}\{\delta(t-a)\} = \int_0^\infty e^{-st}\delta(t-a)dt = e^{-sa} \tag{5.9}$$

$$\mathcal{L}\{\delta'(t-a)\} = \int_0^\infty e^{-st}\delta'(t-a)dt = -\left[\frac{d}{dt}(e^{-st})\right]_{t=a} = se^{-sa} \tag{5.10}$$

and so on.

Now take the case of a function f defined by a relation of the form

$$f(t) = \phi_1(t)u(a-t) + \phi_2(t)u(t-a) \tag{5.11}$$

where $a > 0$, and ϕ_1 and ϕ_2 are continuously differentiable functions. Using the notation suggested in Sec. 3.3 we shall write

and
$$f'(t) = \phi_1'(t)u(a-t) + \phi_2'(t)u(t-a) \quad \text{(for all } t \neq a)$$

$$Df(t) = \phi_1'(t)u(a-t) + \phi_2'(t)u(t-a) + [\phi_2(a) - \phi_1(a)]\,\delta(t-a)$$
$$\equiv f'(t) + [f(a+) - f(a-)]\,\delta(t-a) \,. \tag{5.12}$$

Using integration by parts to evaluate the Laplace integral we have

$$\int_0^\infty e^{-st} f'(t)\,\mathrm{d}t = \int_0^a \phi_1'(t)e^{-st}\,\mathrm{d}t + \int_a^\infty \phi_2'(t)e^{-st}\,\mathrm{d}t$$

$$= \left[e^{-st}\phi_1(t) \right]_0^a + s\int_0^a \phi_1(t)e^{-st}\,\mathrm{d}t + \left[e^{-st}\phi_2(t) \right]_a^\infty + s\int_a^\infty \phi_2(t)e^{-st}\,\mathrm{d}t$$

$$= s\left[\int_0^a \phi_1(t)e^{-st}\,\mathrm{d}t + \int_a^\infty \phi_2(t)e^{-st}\,\mathrm{d}t \right] - e^{-as}\left[\phi_2(a) - \phi_1(a)\right] - \phi_1(0)$$

$$\equiv sF_0(s) - f(0) - e^{-as}\left[f(a+) - f(a-)\right] \tag{5.13}$$

so that a modification of the derivative rule is required when we adhere to the classical meaning of the term "derivative" in the case of discontinuous functions.

On the other hand, from (5.12) we get

$$\int_0^\infty e^{-st}\{Df(t)\}\,\mathrm{d}t = \int_0^\infty e^{-st}f'(t)\,\mathrm{d}t + [f(a+) - f(a-)]\,e^{-as}$$

$$= sF_0(s) - f(0) \tag{5.14}$$

and the usual form of the derivative rule continues to apply.

5.2.2 The result (5.13) makes sense even when we allow a to tend to zero, for then we get

$$\mathcal{L}\{f'(t)\} = \int_0^\infty \phi_2'(t)e^{-st}\,\mathrm{d}t = s\int_0^\infty \phi_2(t)e^{-st}\,\mathrm{d}t - \phi_2(0)$$

$$= sF_0(s) - f(0+) \,. \tag{5.15}$$

However, a complication arises with regard to $\mathcal{L}\{Df(t)\}$ when $a = 0$. If we have

$$f(t) = \phi_1(t)u(-t) + \phi_2(t)u(t)$$

then

$$Df(t) = \phi_1'(t)u(-t) + \phi_2'(t)u(t) + [\phi_2(0) - \phi_1(0)]\,\delta(t)$$

and so,

$$\mathcal{L}\{Df(t)\} = \mathcal{L}\{\phi_2'(t)\} + [\phi_2(0) - \phi_1(0)]\,\mathcal{L}\{\delta(t)\}$$

$$= s\,\mathcal{L}\{\phi_2(t)\} - \phi_2(0) + [\phi_2(0) - \phi_1(0)]\,\Delta(s)$$

$$\equiv sF_0(s) - f(0+) + [f(0+) - f(0-)]\,\Delta(s) . \qquad (5.16)$$

The difficulty is that, as remarked in Sec. 4.5, the Laplace Transform of the delta function (which we have denoted by $\Delta(s)$) is not defined by the Laplace integral

$$\int_0^\infty e^{-st}\delta(t)\,dt = \int_{-\infty}^{+\infty} e^{-st}u(t)\delta(t)\,dt .$$

The role of the delta function as a (generalised) impulse response function suggests that we should have $\Delta(s) = 1$ for all s, and this is the definition most usually adopted. However the discussion given in Sec. 3.2.3 on the significance of the formal product $u(t)\delta(t)$ shows that there are grounds for taking $\Delta(s) = \frac{1}{2}$, for all s; other values for $\Delta(s)$ have also been suggested. It should be noted that the issue cannot be resolved simply by an appeal to the definition of δ as a limit, nor by means of the formulation as a (Riemann) Stieltjes integral. In the latter case, for example, we have for an arbitrary continuous integrand f

$$\int_0^\infty f(t)\,du_c(t) = (1 - c)f(0) \qquad (5.17)$$

(see No. 4 of Exercises III, Chapter 2). We could therefore obtain $\Delta(s) \equiv 1$ by choosing $c = 0$ or, equally well, $\Delta(s) \equiv \frac{1}{2}$ by choosing $c = \frac{1}{2}$. Whatever value we choose for $\Delta(s)$ the relation (5.16) is bound to be consistent with the behaviour of δ as the derivative of the unit step function u. For, since

$$\mathcal{L}\{u(t)\} = \int_0^\infty e^{-st}\,dt = 1/s ,$$

we have

$$\mathcal{L}\{u'(t)\} = \{s\left(\frac{1}{s}\right) - u(0+)\} + \Delta(s)\{u(0+) - u(0-)\}$$

$$= (1 - 1) + \Delta(s)(1 - 0) = \Delta(s) .$$

On the other hand care must be taken to ensure that the correct form of (5.16) is used when a specific definition of $\Delta(s)$ has been decided on. Thus, for $\Delta(s) = 1$ we get

$$\mathcal{L}\{Df(t)\} = sF_0(s) - f(0-) \tag{5.18}$$

$$= sF_0(s)$$

whenever $f(t) = 0$ for all $t < 0$.

But for $\Delta(s) = \frac{1}{2}$ the result becomes

$$\mathcal{L}\{Df(t)\} = sF_0(s) - \frac{1}{2}\{f(0+) + f(0-)\} .$$

In what follows we shall adopt the majority view and define $\Delta(s)$ to be 1 for all values of s. Similarly we shall take the Laplace Transform of δ' to be s; the analogue of (5.18) then becomes

$$\mathcal{L}\{D^2f(t)\} = s^2F_0(s) - sf(0-) - f'(0-) \tag{5.19}$$

$$= s^2F_0(s)$$

whenever $f(t) = 0$ for all $t < 0$. The convenience of these definitions is readily illustrated by the following derivation of the Laplace Transform of a **periodic function**:

Let f be a function which vanishes identically outside the finite interval $(0,T)$. The **periodic extension** of f, of period T, is the function obtained by summing the translates, $f(t - kT)$, for $k = 0, \pm 1, \pm 2, \dots$, (see Fig. 5.1)

$$f_T(t) = \sum_{k=-\infty}^{+\infty} f(t - kT) \tag{5.20}$$

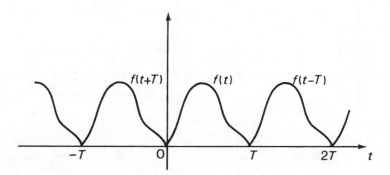

Fig. 5.1.

We can write f_T as a convolution:

$$f_T(t) = \sum_{k=-\infty}^{+\infty} \{f(t) * \delta(t - kT)\} = f(t) * \sum_{k=-\infty}^{+\infty} \delta(t - kT). \qquad (5.21)$$

Further, using the above definition of $\Delta(s)$, we obtain

$$\mathcal{L} \left[\sum_{k=-\infty}^{+\infty} \delta(t - kT) \right] = \mathcal{L} \left[\sum_{k=0}^{\infty} \delta(t - kT) \right]$$

$$= 1 + e^{-sT} + e^{-2sT} + e^{-3sT} + \ldots = \frac{1}{1 - e^{-sT}} \qquad (5.22)$$

the summation being valid provided that

$$|e^{-sT}| = |e^{-(\sigma + i\omega)T}| = e^{-\sigma T} < 1 ,$$

that is, for all s such that $\mathrm{Re}(s) > 0$. Hence, appealing to the Convolution Theorem for the Laplace Transform, (5.21) and (5.22) together yield

$$\mathcal{L} \left[\sum_{k=-\infty}^{+\infty} f(t - kT) \right] = \frac{F_0(s)}{1 - e^{-sT}} . \qquad (5.23)$$

Exercises II

1. Identify the (generalised) functions whose Laplace Transforms are:

    (a) $\dfrac{s^3 + 2}{s + 1}$; (b) $\dfrac{\cosh s}{e^s}$; (c) $\dfrac{1 - s^n e^{-ns}}{1 - se^{-s}}$.

2. Find the Laplace Transforms of the periodic functions sketched in Fig. 5.2.

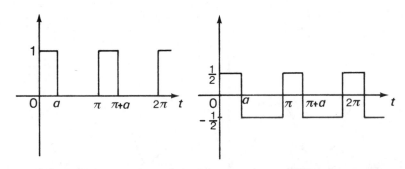

Fig. 5.2 (a).   Fig. 5.2 (b).

3. Show that $\mathcal{L}\{\delta(\cos t)\} = e^{-\pi s/2}\big/[1 - e^{-\pi s}]$.

5.3 COMPUTATION OF LAPLACE TRANSFORMS

5.3.1 If f is an ordinary function whose Laplace Transform exists (for some values of s) then we should be able to find that transform, in principle at least, by evaluating directly the integral which defines $F_0(s)$. It is usually simpler in practice to make use of certain appropriate properties of the Laplace integral and to derive specific transforms from them. The following results are easy to establish and are particularly useful in this respect:

(L.T.1) *The First Translation Property.* If $\mathcal{L}\{f(t)\} = F_0(s)$, and if a is any real constant, then

$$\mathcal{L}\{e^{at}f(t)\} = F_0(s-a) .$$

(L.T.2) *The Second Translation Property.* If $\mathcal{L}\{f(t)\} = F_0(s)$, and if a is any positive constant, then

$$\mathcal{L}\{u(t-a)f(t-a)\} = e^{-as}F_0(s) .$$

(L.T.3) *Change of Scale.* If $\mathcal{L}\{f(t)\} = F_0(s)$, and if a is any positive constant, then

$$\mathcal{L}\{f(at)\} = \frac{1}{a}F_0\left(\frac{s}{a}\right).$$

(L.T.4) *Multiplication by t.* If $\mathcal{L}\{f(t)\} = F_0(s)$, then

$$\mathcal{L}\{tf(t)\} = -\frac{\mathrm{d}}{\mathrm{d}s}F_0(s) \equiv -F_0'(s) .$$

(L.T.5) *Transform of an Integral.* If $\mathcal{L}\{f(t)\} = F_0(s)$, and if the function g is defined by

$$g(t) = \int_0^t f(\tau)\mathrm{d}\tau$$

then

$$\mathcal{L}\{g(t)\} = \frac{1}{s}F_0(s) .$$

The first three of the above properties follow immediately on making suitable changes of variable in the Laplace integrals concerned. For (L.T.4) we have only to differentiate with respect to s under the integral sign, while in the case of (L.T.5) it is enough to note that $g'(t) = f(t)$ and that $g(0) = 0$; the result then follows from the rule for finding the Laplace Transform of a derivative. Using these properties, an elementary basic table of standard transforms can be constructed without difficulty (Table I). This list can be extended by using various special techniques. In particular the results for the transforms of delta functions derived in the preceding section are of considerable value in this connection.

Table I

$f(t)$	$F_0(s)$	Region of (absolute) convergence
$u(t)$	$1/s$	$\mathrm{Re}(s) > 0$
t	$1/s^2$	$\mathrm{Re}(s) > 0$
$t^n \ (n > 1)$	$n!/s^{n+1}$	$\mathrm{Re}(s) > 0$
e^{at}	$\dfrac{1}{s-a}$	$\mathrm{Re}(s) > a$
e^{-at}	$\dfrac{1}{s+a}$	$\mathrm{Re}(s) > -a$
$\sinh at$	$\dfrac{a}{s^2 - a^2}$	$\mathrm{Re}(s) > |a|$
$\cosh at$	$\dfrac{s}{s^2 - a^2}$	$\mathrm{Re}(s) > |a|$
$\sin at$	$\dfrac{a}{s^2 + a^2}$	$\mathrm{Re}(s) > 0$
$\cos at$	$\dfrac{s}{s^2 + a^2}$	$\mathrm{Re}(s) > 0$

5.3.2

Example 1. Find the Laplace transform of the triangular waveform shown in Fig. 5.3.

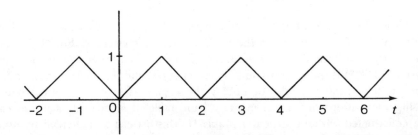

Fig. 5.3.

We shall obviously expect to use the formula (5.23) for the Laplace Transform of the periodic extension of a function f, but the first need is to establish the transform of this function f itself. In Fig. 5.4 there is shown a decomposition of the required function into a combination of ramp functions:

$$f(t) = tu(t) - 2(t-1)u(t-1) + (t-2)u(t-2)$$

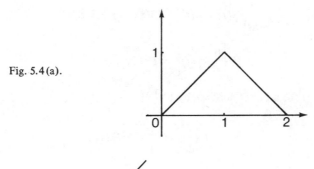

Fig. 5.4 (a).

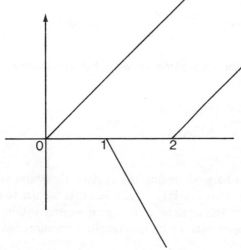

Fig. 5.4 (b).

A straightforward application of the second translation property (L.T.2) immediately gives

$$F_0(s) = \frac{1}{s^2} - \frac{2}{s^2}e^{-s} + \frac{e^{-2s}}{s^2} = \left[\frac{1-e^{-s}}{s}\right]^2 = \frac{4}{s^2}e^{-s}\sinh^2\frac{s}{2}.$$

Hence, applying (5.23),

$$\mathcal{L}\{f_T(t)\} = \left[\frac{4}{s^2}e^{-s}\sinh^2\frac{s}{2}\right]\left[\frac{1}{1-e^{-2s}}\right] = \frac{2\sinh^2 s/2}{s^2 \sinh s} = \frac{\tanh s/2}{s^2}.$$

However, brief as this calculation is, it is worthwhile examining an alternative attack which relies on the use of the delta functions and of the differentiation formula (5.19). We have only to differentiate f twice to eliminate all functions other than delta functions:

$$\frac{d}{dt}f(t) = u(t) - u(t-1) + u(t-2)$$

$$\frac{d^2}{dt^2}f(t) = \delta(t) - 2\delta(t-1) + \delta(t-2).$$

Hence,

$$\mathcal{L}\left[\frac{d^2}{dt^2}f(t)\right] = 1 - 2e^{-s} + e^{-2s}.$$

But $f(t) = 0$ for all $t < 0$ and so, by (5.19),

$$\mathcal{L}\left[\frac{d^2}{dt^2}f(t)\right] = s^2 F_0(s).$$

That is,

$$F_0(s) = \frac{1}{s^2}\mathcal{L}\left[\frac{d^2}{dt^2}f(t)\right] = \frac{1 - 2e^{-s} + e^{-2s}}{s^2} = \frac{4e^{-s}}{s^2}\sinh^2 s/2.$$

Using equation (5.23), the triangular waveform shown in Fig. 5.3 is seen to have the following Laplace Transform:

$$\frac{4e^{-s}\sinh^2 s/2}{s^2(1-e^{-2s})} = \frac{2\sinh^2 s/2}{s^2\sinh s} = \frac{1}{s^2}\tanh\frac{s}{2}.$$

5.3.3 A similar approach could be used to find the Laplace Transform of the alternating triangular waveform shown in Fig. 5.5. In fact it is simpler to note that such a periodic function, exhibiting skew, or mirror, symmetry within each period, can be represented as a convolution with an alternating impulse train:

$$f(t) * \sum_{k=-\infty}^{+\infty} (-1)^k \delta(t - kT/2) \tag{5.24}$$

where f now denotes the pulse shape in the first half-period. The transform of the impulse train is given by

$$\mathcal{L}\left[\sum_{k=-\infty}^{+\infty}(-1)^k\delta\left(t - \frac{kT}{2}\right)\right] = \mathcal{L}\left[\sum_{k=0}^{\infty}(-1)^k\delta(t - kT/2)\right] = \frac{1}{1 + e^{-sT/2}}. \tag{5.25}$$

Accordingly the Laplace Transform of (5.24) is given, by the Convolution Theorem, as

$$\frac{F_0(s)}{1 + e^{-sT/2}}.\tag{5.26}$$

In particular, for the function shown in Fig. 5.5 we have the transform

$$\frac{4e^{-s}}{s^2} \frac{\sinh^2 s/2}{1 + e^{-2s}} = \frac{2 \sinh^2 s/2}{s^2 \cosh s}.$$

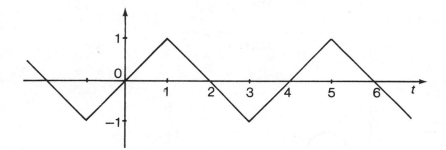

Fig. 5.5.

The technique used to compute $\mathcal{L}\{f(t)\}$ above can obviously be generalised. Let f be a function which vanishes for all $t < 0$ and whose graph consists of a finite number of arcs each defined by a polynomial. Then, for some value of n, the nth derivative $\dfrac{d^n}{dt^n} f(t)$ will consist only of delta functions and derivatives of delta functions; this sequence of differentiations can often be done by inspection. The Laplace Transform of f can then be written down by using the appropriate generalisation of (5.19):

$$\mathcal{L}\left[\frac{d^n}{dt^n} f(t)\right] = s^n F_0(s).$$

5.3.4 Variants of the method can be used in certain situations where f does not reduce to delta function form after finitely many differentiations, as in the following example:

Example 2. Let $f(t) = \sin t$ for $0 < t < \pi$, and $f(t) = 0$ for all other values of t. Then,

$$\frac{d}{dt} f(t) = \cos t \left[u(t) - u(t - \pi)\right]$$

and

$$\frac{\mathrm{d}^2}{\mathrm{d}t^2} f(t) = -\sin t \, [u(t) - u(t - \pi)] + \delta(t) + \delta(t - \pi)$$

$$= \delta(t) + \delta(t - \pi) - f(t) \, .$$

(These results can certainly be obtained by inspection, as a glance at Fig. 5.6 should make clear.) Thus we can write

$$s^2 \, F_0(s) \equiv \mathcal{L}\left[\frac{\mathrm{d}^2}{\mathrm{d}t^2} f(t)\right] = 1 + e^{-\pi s} - F_0(s)$$

so that,

$$F_0(s) = \frac{1 + e^{-\pi s}}{s^2 + 1} \, .$$

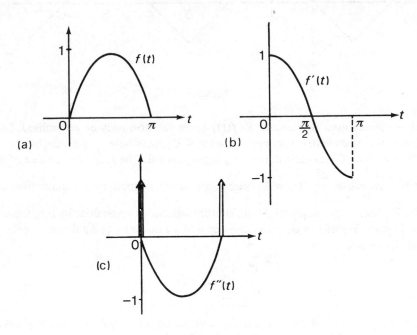

Fig. 5.6.

Applying (5.23) we obtain at once the Laplace Transforms of the periodic waveforms shown in Fig. 5.7 ("rectified" sine waves):

$$\mathcal{L}\{f_{2\pi}(t)\} = \frac{1 + e^{-\pi s}}{s^2 + 1} \frac{1}{1 - e^{-2\pi s}} = \frac{e^{\pi s/2}}{2(s^2 + 1) \sinh \pi s/2}$$

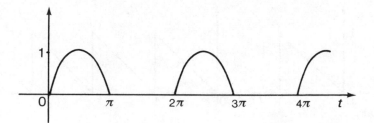

Fig. 5.7 (a).

$$\mathcal{L}\{f_\pi(t)\} = \frac{1+e^{-\pi s}}{s^2+1} \frac{1}{1-e^{-\pi s}} = \frac{\coth \pi s/2}{s^2+1}$$

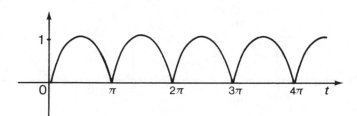

Fig. 5.7 (b).

Exercises III

1. Find the Laplace Transforms of each of the following functions:

    (a) $\cosh^2 at$; (b) $e^{-|t|} \sin^2 t$; (c) $t^3 \cosh t$; (d) $t \cos at$;
    (e) $t^2 \sin at$.

2. If $\mathcal{L}\{f(t)\} \equiv F_0(s)$, show that $\mathcal{L}\{f(t)/t\} = \displaystyle\int_s^\infty F_0(p)dp$, provided that

    $\lim_{t \to 0} f(t)/t$ exists. Hence find the transforms of

    (a) $(\sin t)/t$; (b) $\{\cos at - \cos bt\}/t$; (c) $\displaystyle\int_0^t \frac{\sin \tau \, d\tau}{\tau}$.

3. Find the Laplace Transforms of each of the functions sketched in Fig.
    5.8(a)–(g).

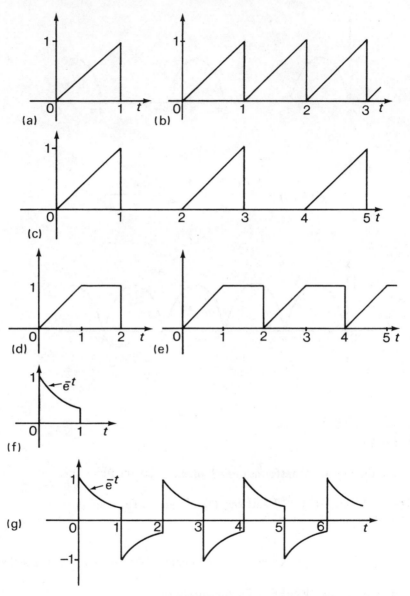

Fig. 5.8.

5.4 NOTE ON INVERSION

5.4.1 There is an inversion theorem for the Laplace Transform which expresses $f(t)$ explicitly in terms of $F_0(s)$. In spite of the generality of the result we shall not enlarge on it here, partly because it presupposes some acquaintance with

Complex Variable Theory, and partly because its use is not really necessary, or even desirable, in many elementary applications. Accordingly we confine attention here to a brief account of those techniques likely to be of most value in the inversion of commonly occurring types of transforms.

Suppose first that $F_0(s) = N(s)/D(s)$, where N and D are polynomials in s with real coefficients. Without loss of generality we may assume that the degree of N is less than the degree of D; for otherwise we can divide by $D(s)$ to obtain

$$F_0(s) = P(s) + \frac{N_0(s)}{D(s)}$$

where P is a polynomial in s, and the degree of the polynomial N_0 is less than that of D. Since $\mathcal{L}\{\delta^{(n)}(t)\} = s^n$, we can at once identify P as the transform of a linear combination of delta functions. Similarly, there is no loss of generality in assuming that $D(s)$ is normalised so that the coefficient of the highest power of s, say s^m, occurring in the denominator of $F_0(s)$ is unity. Then we may write, in general,

$$D(s) = (s - \alpha_1)^{p_1} (s - \alpha_2)^{p_2} \ldots (s - \alpha_m)^{p_m} \tag{5.27}$$

where the zeros, α_k, may be real or complex.

5.4.2 If the zeros of $D(s)$ are simple, so that in (5.27) we have

$$p_1 = p_2 = \ldots = p_m = 1$$

then the decomposition of $F_0(s)$ into partial fractions takes the form

$$F_0(s) \equiv \frac{N(s)}{D(s)} = \frac{A_1}{s - \alpha_1} + \frac{A_2}{s - \alpha_2} + \ldots + \frac{A_m}{s - \alpha_m}.$$

The numbers A_k may be determined either by equating coefficients of powers of s, or by substituting particular values for s so as to generate a system of linear algebraic equations in the A_k. Alternatively, note that

$$(s - \alpha_1)\frac{N(s)}{D(s)} = A_1 + A_2 \frac{s - \alpha_1}{s - \alpha_2} + \ldots + A_m \frac{s - \alpha_1}{s - \alpha_m}$$

$$(s - \alpha_2)\frac{N(s)}{D(s)} = A_1 \frac{s - \alpha_1}{s - \alpha_2} + A_2 + \ldots + A_m \frac{s - \alpha_2}{s - \alpha_m}$$

and so on. It follows that

$$A_k = \lim_{s \to \alpha_k} \frac{(s - \alpha_k)N(s)}{D(s)}, \quad k = 1, 2, \ldots, m.$$

Using l'Hopital's rule for the evaluation of the limit gives the following explicit expression for A_k:

$$A_k = \lim_{s \to \alpha_k} \left[\frac{N(s) + (s - \alpha_k)N'(s)}{D(s)} \right] = \frac{N(\alpha_k)}{D'(\alpha_k)} .$$

Since $\mathcal{L}\{e^{\alpha t}\} = 1/(s - \alpha)$, this means that we can write down the inverse transform of $F_0(s)$ in the form

$$f(t) = \frac{N(\alpha_1)}{D'(\alpha_1)} e^{\alpha_1 t} + \frac{N(\alpha_2)}{D'(\alpha_2)} e^{\alpha_2 t} + \ldots + \frac{N(\alpha_m)}{D'(\alpha_m)} e^{\alpha_m t} . \tag{5.28}$$

The expression (5.28) applies whether the zeros of $D(s)$ turn out to be real or complex. However, since $D(s)$ is a polynomial with real coefficients, any complex zeros must necessarily occur in complex conjugate pairs; that is, $D(s)$ will contain, at worst, real quadratic factors of the form $s^2 + bs + c$, where $b^2 < 4c$. To rewrite (5.28) in a more compact form, free from complex exponentials, usually involves a good deal of algebraic manipulation. It is generally easier to make use directly of the existence of real quadratic factors of $D(s)$; in the partial fraction expansion such a quadratic factor gives rise to expressions like

$$\frac{As + B}{s^2 + bs + c} .$$

Now we can always write

$$s^2 + bs + c = (s + \alpha)^2 + \beta^2$$

where

$$\alpha = b/2 \quad \text{and} \quad \beta^2 = \frac{1}{4}(4c - b^2) .$$

Hence,

$$\frac{As + B}{s^2 + bs + c} = \frac{As + B}{(s + \alpha)^2 + \beta^2} = \frac{A(s + \alpha)}{(s + \alpha)^2 + \beta^2} + \frac{B - \alpha A}{(s + \alpha)^2 + \beta^2} .$$

Then, using the Second Translation Property (L.T.2), and recalling the standard transforms of $\sin at$ and $\cos at$, we can write down the required inverse transform as

$$A e^{-\alpha t} \cos \beta t + \frac{(B - \alpha A)}{\beta} e^{-\alpha t} \sin \beta t . \tag{5.29}$$

5.4.3 Suppose, on the other hand, that $D(s)$ has a repeated factor, say $(s - \alpha)^2$. This time the partial fraction expansion will generally include the terms

$$\frac{A_{11}}{(s - \alpha)^2} + \frac{A_{12}}{(s - \alpha)} .$$

To determine A_{11} and A_{12}, note that

$$A_{11} = \lim_{s \to \alpha} \left[(s - \alpha)^2 \frac{N(s)}{D(s)} \right]; \quad A_{12} = \lim_{s \to \alpha} \left[\frac{d}{ds} (s - \alpha)^2 \frac{N(s)}{D(s)} \right].$$

Similarly, if $D(s)$ has a factor $(s - \alpha)^3$ then the partial fraction expansion will include the terms

$$\frac{A_{11}}{(s - \alpha)^3} + \frac{A_{12}}{(s - \alpha)^2} + \frac{A_{13}}{s - \alpha}$$

and we can determine A_{11}, A_{12}, and A_{13} from

$$A_{11} = \lim_{s \to \alpha} \left[(s - \alpha)^3 \frac{N(s)}{D(s)} \right]$$

$$A_{12} = \lim_{s \to \alpha} \left[\frac{d}{ds} (s - \alpha)^3 \frac{N(s)}{D(s)} \right]$$

$$A_{13} = \lim_{s \to \alpha} \left[\frac{1}{2} \frac{d^2}{ds^2} (s - \alpha)^3 \frac{N(s)}{D(s)} \right].$$

To invert the terms we need only appeal to the standard result

$$\mathcal{L}\{t^{k-1} e^{\alpha t}\} = (k - 1)! \left[\frac{1}{(s - \alpha)^k} \right].$$

In principle, similar methods could be used whenever $D(s)$ has a factor $(s - \alpha)^n$, for any positive integer n; however, for $n \geqslant 3$, the successive differentiations required to determine the partial fraction coefficients as above make the work prohibitive.

5.4.4 For transforms which are not rational functions of s, more high-powered methods are usually required. However, for some types of transcendental functions simpler techniques suffice. In particular, recall that the presence of a pure exponential factor, e^{-as}, in the transform corresponds to a delay, of magnitude a, in the inverse transform. Similarly a factor of the form $(1 - e^{-sT})^{-1}$, or $(1 + e^{-sT})^{-1}$, indicates that the inverse transform is a convolution with a periodic impulse train as one of the components; the resulting function may or may not be periodic. For example, it is easy to invert the following two transforms after a little elementary manipulation:

$$F_0(s) = \frac{1}{s \cosh s}; \quad G_0(s) = \frac{1}{s \sinh s}.$$

We have:

$$F_0(s) = \frac{1}{s} \frac{2}{e^s + e^{-s}} = \frac{2}{se^s} \frac{1}{1 + e^{-2s}} = \frac{2e^{-s}}{s} \{1 - e^{-2s} + e^{-4s} - \ldots\}$$

$$G_0(s) = \frac{1}{s} \frac{2}{e^s - e^{-s}} = \frac{2}{se^s} \frac{1}{1 - e^{-2s}} = \frac{2e^{-s}}{s} \{1 + e^{-2s} + e^{-4s} + \ldots\}$$

By inspection we can identify the required inverse transforms as the functions shown in Fig. 5.9(a) and Fig. 5.9(b) respectively.

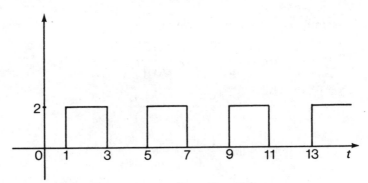

Fig. 5.9 (a).

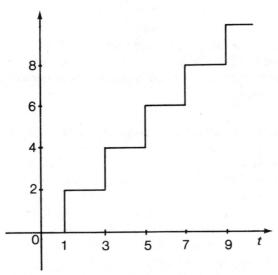

Fig. 5.9 (b).

CHAPTER 6

Fourier Series and Fourier Transforms

6.1 FOURIER SERIES

6.1.1 The classical theory of the Fourier series expansion of a periodic function can be summarised as follows.

Let f be a real function of t, defined for all t such that $-T/2 < t < +T/2$, and write $\omega_0 \equiv 2\pi/T$. Assume that at all points of the open interval at which f is continuous the following expansion is valid:

$$f(t) = \frac{a_0}{2} + \sum_{n=1}^{\infty} \{a_n \cos n\omega_0 t + b_n \sin n\omega_0 t\} . \qquad (6.1)$$

We can obtain explicit formulas for the coefficients a_0, a_n, and b_n appearing in (6.1) by using the so-called **orthogonality relations** for the sine and cosine functions,

$$\int_{-T/2}^{+T/2} \cos n\omega_0 t \cos m\omega_0 t \, dt = \begin{cases} T/2 & \text{if } m = n \\ 0 & \text{if } m \neq n \end{cases}$$

$$\int_{-T/2}^{+T/2} \sin n\omega_0 t \cos m\omega_0 t \, dt = 0 \qquad \text{for all } m \text{ and } n,$$

$$\int_{-T/2}^{+T/2} \sin n\omega_0 t \sin m\omega_0 t \, dt = \begin{cases} T/2 & \text{if } m = n \\ 0 & \text{if } m \neq n , \end{cases}$$

Thus, for any given integer m, we can multiply (6.1) throughout by $\cos m\omega_0 t$ and integrate from $-T/2$ to $T/2$ to get

$$\int_{-T/2}^{+T/2} f(t) \cos m\omega_0 t \, dt = \int_{-T/2}^{+T/2} \cos m\omega_0 t \left[\frac{a_0}{2} + \sum_{n=1}^{\infty} \{a_n \cos n\omega_0 t + b_n \sin n\omega_0 t\} \right] dt .$$

Rearranging the terms on the right-hand side, we can write this equation in the following form

$$\int_{-T/2}^{+T/2} f(t)\cos m\omega_0 t\, dt = \frac{a_0}{2}\int_{-T/2}^{+T/2}\cos m\omega_0 t\, dt$$

$$+\sum_{n=1}^{\infty}\left[\int_{-T/2}^{+T/2}[a_n\cos m\omega_0 t\cos n\omega_0 t + b_n\cos m\omega_0 t\sin n\omega_0 t]\, dt\right]$$

always provided that it is permissible to integrate term by term on the right-hand side. Then all the terms on the terms on the right-hand side vanish except when $m = n$, and we can solve for the coefficient a_m. Repeating the process with $\sin m\omega_0 t$ instead of $\cos m\omega_0 t$ we can similarly obtain the coefficient b_m. In the event the required formulas turn out to be

$$a_n = \frac{2}{T}\int_{-T/2}^{+T/2} f(t)\cos n\omega_0 t\, dt\; ; \quad b_n = \frac{2}{T}\int_{-T/2}^{+T/2} f(t)\sin n\omega_0 t\, dt$$

for $n = 1, 2, 3, \ldots$, and   (6.2)

$$a_0 = \frac{2}{T}\int_{-T/2}^{+T/2} f(t)\, dt\;.$$

In point of fact it is not generally true that, for an arbitrary function f, the **Fourier series** given in the right-hand side of (6.1) will converge at each particular point t to the value $f(t)$ assumed by the function f at that point. To ensure that convergence does occur it is necessary to impose extra constraints on f. For our purposes it will be enough to assume that f satisfies the **Dirichlet conditions** given below, although these are rather more stringent than necessary:

(i) f is bounded on the interval $(-T/2, +T/2)$,
(ii) the interval $(-T/2, +T/2)$ may be divided into a finite number of sub-intervals in each of which the derivative f' exists throughout and does not change sign.

With these conditions fulfilled it is certainly the case that (6.1) is valid at each point of continuity of f in $(-T/2, +T/2)$. Moreover, at any point t in the interval at which f is discontinuous, the one-sided limits $f(t+)$ and $f(t-)$ both exist and we have

$$\frac{1}{2}\{f(t+) + f(t-)\} = \frac{a_0}{2} + \sum_{n=1}^{\infty}\{a_n\cos n\omega_0 t + b_n\sin n\omega_0 t\}\;.\qquad(6.3)$$

We have assumed so far that f is a function defined only on the fundamental interval $(-T/2, +T/2)$ and have tacitly considered the convergence of the Fourier series of f only at points of this interval. In fact if the series on the right-hand side of (6.1) converges at some point t_0 in $(-T/2, +T/2)$ then it will also converge, and to the same value, at every point of the form $t = t_0 + nT$ where $n = \pm 1, \pm 2, \pm 3, \ldots$. If we denote by f_T the function obtained by periodically repeating f, with period T, then we can express this fact by writing

$$f_T(t) \cong \sum_{n=-\infty}^{+\infty} f(t - nT) \cong \frac{a_0}{2} + \sum_{n=1}^{\infty} \{a_n \cos n\omega_0 t + b_n \sin n\omega_0 t\} \qquad (6.4)$$

where the symbol "$\cong$" is taken to mean equality if t is a point of continuity of f and is to be interpreted in the sense of equation (6.3) otherwise.

6.1.2 It is often convenient to write (6.1) and (6.2) in the following alternative form:

$$f(t) \cong \frac{A_0}{T} + \frac{2}{T} \sum_{n=1}^{\infty} \{A_n \cos n\omega_0 t + B_n \sin n\omega_0 t\} \qquad (6.5)$$

where

$$A_n = \frac{Ta_n}{2} = \int_{-T/2}^{+T/2} f(t) \cos n\omega_0 t \, dt \; ; \quad B_n = \frac{Tb_n}{2} = \int_{-T/2}^{+T/2} f(t) \sin n\omega_0 t \, dt$$

for $n = 1, 2, 3, \ldots$, and

$$A_0 = \frac{Ta_0}{2} = \int_{-T/2}^{+T/2} f(t) \, dt.$$

Using this form, we can easily derive the so-called **exponential form** of the Fourier series of f:

$$f(t) \cong \frac{A_0}{T} + \frac{1}{T} \sum_{n=1}^{\infty} \{A_n(e^{in\omega_0 t} + e^{-in\omega_0 t}) + \frac{B}{i}(e^{in\omega_0 t} - e^{-in\omega_0 t})\}$$

$$= \frac{A_0}{T} + \frac{1}{T} \sum_{n=1}^{\infty} \{(A_n - iB_n) e^{in\omega_0 t} + (A_n + iB_n) e^{-in\omega_0 t}\}$$

$$= \frac{1}{T} \sum_{n=-\infty}^{+\infty} C_n e^{in\omega_0 t} \qquad (6.6)$$

where

$$C_n = A_n - iB_n = \int_{-T/2}^{+T/2} f(t)\, e^{-in\omega_0 t} dt$$

and

$$C_{-n} = A_n + iB_n = \bar{C}_n . \tag{6.7}$$

(the bar denoting the complex conjugate).

Note that we could have obtained the formulas (6.7) directly by assuming the expansion in the form (6.6) in the first place, multiplying throughout by $\exp(-im\omega_0 t)$ for an arbitrary fixed integer m, and then integrating from $-T/2$ to $T/2$ to determine C_m.

Exercises I

1. If f satisfies the Dirichlet conditions in $-T/2 < t < T/2$, and is an *even* function, show that its Fourier series consists entirely of cosine terms; similarly, if f is an *odd* function, show that its Fourier series consists entirely of sine terms.

2. Obtain Fourier expansions valid in the interval $-\pi < t < +\pi$ for each of the following:

    (a) $f_1(t) = t$;   (b) $f_2(t) = |t|$;   (c) $f_3(t) = t^2$.

    By choosing a suitable value of t in each case deduce that

    (d) $\pi/4 = 1 - \dfrac{1}{3} + \dfrac{1}{5} - \dfrac{1}{7} + \ldots$

    (e) $\pi^2/8 = 1 + \dfrac{1}{3^2} + \dfrac{1}{5^2} + \dfrac{1}{7^2} + \ldots$

    (f) $\pi^2/12 = 1 - \dfrac{1}{2^2} + \dfrac{1}{3^2} - \dfrac{1}{4^2} + \ldots$

3. Obtain a Fourier expansion, valid for $-\pi < t < +\pi$, for the function $\cos xt$ where x is some fixed number which is *not* an integer. By letting $t \to \pi$ deduce that

    $$\cot \pi x - \frac{1}{\pi x} = -\frac{2x}{\pi}\left[\frac{1}{1^2 - x^2} + \frac{1}{2^2 - x^2} + \frac{1}{3^2 - x^2} + \ldots\right]$$

4. (a) If $f_1(t) = \cos t$ for $0 < t < \pi$, expand f_1 as a Fourier series, valid at least in $0 < t < \pi$, which consists wholly of sine terms. [Hint: extend f_1 as an *odd* function on $-\pi < t < +\pi$.]

    (b) If $f_2(t) = \sin t$ for $0 < t < \pi$, expand f_2 as a Fourier series, valid at least in $0 < t < \pi$, which consists wholly of cosine terms. [Hint: extend f_2 as an *even* function on $-\pi < t < +\pi$.]

6.2 GENERALISED FOURIER SERIES

6.2.1 If we allow a purely formal application of the results of the preceding sections then we can derive very simply a Fourier series representation of a periodic train of delta functions,

$$\sum_{n=-\infty}^{+\infty} \delta(t-nT) .$$

Thus, if we assume that

$$\sum_{n=-\infty}^{+\infty} \delta(t-nT) = \frac{1}{T} \sum_{n=-\infty}^{+\infty} C_n \, e^{in\omega_0 t}$$

then the coefficients C_n would appear to be given by

$$C_n = \int_{-T/2}^{+T/2} \delta(t)\, e^{-in\omega_0 t} \mathrm{d}t = \left[e^{-in\omega_0 t} \right]_{t=0} = 1 , \quad \text{for all } n.$$

Hence, again purely formally, we obtain

$$\sum_{n=-\infty}^{+\infty} \delta(t-nT) = \frac{1}{T} \sum_{n=-\infty}^{+\infty} e^{in\omega_0 t} = \frac{1}{T} \left[1 + 2 \sum_{n=1}^{\infty} \cos n\omega_0 t \right]. \tag{6.8}$$

The trigonometric and exponential expressions appearing in (6.8) are divergent. This makes it clear that they are to be interpreted symbolically, and that (6.8) itself is to be understood in the sense that each of the three given expressions is to have the same operational significance.

6.2.2 Some idea of the practical uses to which such symbolic expressions can be put is given by the following illustration of the computation of Fourier coefficients. Consider the periodic function shown in Fig. 6.1.

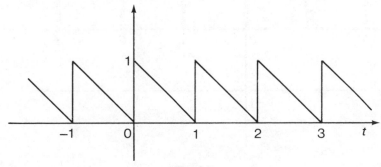

Fig. 6.1.

This has period $T = 1$, and if we assume that

$$f_T(t) = \frac{1}{T} \sum_{n=-\infty}^{+\infty} C_n e^{i2\pi nt}$$

then the coefficients C_n would be given by the integral formula

$$C_n = \int_{-1/2}^{+1/2} f(t) e^{-i2\pi nt} dt = \int_0^1 (1-t) e^{-i2\pi nt} dt . \qquad (6.9)$$

This integral is not particularly difficult to evaluate, but it is worth noting that we can avoid integration altogether if we appeal to (6.8). We have

$$\frac{d}{dt} f_T(t) = \frac{1}{T} \sum_{n=-\infty}^{+\infty} i2\pi n C_n e^{i2\pi nt} . \qquad (6.9)$$

But, as a glance at Fig. 6.1 and Fig. 6.2 shows, the derivative of $f_T(t)$ is given by

$$\frac{d}{dt} f_T(t) = \sum_{n=-\infty}^{+\infty} \delta(t-n) - 1$$

$$= \frac{1}{T} \sum_{n=-\infty}^{+\infty} e^{i2\pi nt} - 1 . \qquad (6.10)$$

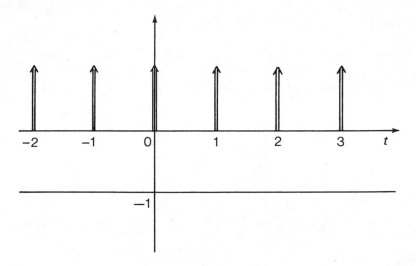

Fig. 6.2.

For any $n \neq 0$ we can equate coefficients of $e^{i2\pi nt}$ in (6.9) and (6.10) to obtain

$$i2\pi nC_n = 1, \quad \text{or} \quad C_n = 1/i2\pi n .$$

C_0 is seen immediately to be 1/2, since the value of the integral in question is simply the area of the triangle formed by the graph of $f(t)$. Thus, in a particularly simple and straightforward manner we get the Fourier series expansion as

$$f_T(t) \cong \frac{1}{2} + \sum_{\substack{n=-\infty \\ n \neq 0}}^{+\infty} \frac{e^{i2\pi nt}}{i2n\pi} = \frac{1}{2} + \sum_{n=1}^{\infty} \frac{1}{n\pi} \sin 2\pi nt .$$

Exercises II

1. Let $f(t) = k$ for $-d/2 < t < d/2$, and $f(t) = 0$ otherwise. Express the periodic extension f_T of f, where $T \geqslant d$, as a Fourier series. Examine the particular cases $d = T, d = T/2$.

    By taking $k = 1/d$ and allowing d to tend to zero, confirm that the formal Fourier series expansion of a periodic train of delta functions is as given in equation (6.8).

2. Obtain (symbolic) Fourier series for each of the following:

    (a) $\displaystyle\sum_{n=-\infty}^{+\infty} \delta(t - 2n\pi)$; (b) $\displaystyle\sum_{n=-\infty}^{+\infty} \delta\{t - (2n + 1)\pi\}$;

    (c) $\displaystyle\sum_{n=-\infty}^{+\infty} (-1)^n \, \delta(t - n\pi)$.

3. Show how the results of Question 2 above could be used to derive the Fourier series expansions over $(-\pi, +\pi)$ for the functions f_1, f_2, and f_3 of Question 2 of Exercises I.

4. Express $\displaystyle\sum_{n=-\infty}^{+\infty} \delta\left[t - (2n + 1)\frac{\pi}{2}\right]$ as a formal Fourier series, and use this

    result to show that

    $$|\cos t| = \frac{2}{\pi} + \frac{4}{\pi} \left[\frac{\cos 2t}{1.3} - \frac{\cos 4t}{3.5} + \frac{\cos 6t}{5.7} - \cdots \right]$$

    (Hint: recall the method used to compute $\mathcal{L}\{|\sin t|\}$ in Sec. 5.3.4).

6.3 FOURIER TRANSFORMS

6.3.1 The full significance and value of the delta function in Fourier analysis can be more readily appreciated if we adopt a somewhat different approach to the study of Fourier series itself. First, let f be a complex-valued function of the real variable t which is *absolutely integrable* over the whole interval $-\infty < t < +\infty$. That is to say we have

$$f(t) = f_1(t) + if_2(t)$$

where

$$\int_{-\infty}^{+\infty} |f_1(t)| \, dt < +\infty \quad \text{and} \quad \int_{-\infty}^{+\infty} |f_2(t)| \, dt < +\infty .$$

Then we define the **Fourier Transform** of f to be the function $F(i\omega)$ defined by

$$F(i\omega) = \int_{-\infty}^{+\infty} e^{-i\omega t} f(t) \, dt . \tag{6.11}$$

The condition of absolute integrability on f is enough to ensure not merely that the function F is well-defined for all ω but that it is actually bounded and everywhere continuous. We shall sometimes write

$$F(i\omega) \equiv \mathcal{F}\{f(t)\} .$$

$F(i\omega)$ is generally a complex-valued function of ω. In case f is itself a real function of t we have

$$F(i\omega) = \int_{-\infty}^{+\infty} e^{-i\omega t} f(t) \, dt = A(\omega) - iB(\omega) \tag{6.12}$$

where the real and imaginary parts of $F(i\omega)$ are given by

$$A(\omega) = \int_{-\infty}^{+\infty} f(t) \cos \omega t \, dt , \quad B(\omega) = \int_{-\infty}^{+\infty} f(t) \sin \omega t \, dt \tag{6.13}$$

A is clearly an even function of ω and is often called the **cosine transform** of f; similarly B is an odd function of ω which is often called the **sine** transform of f. (Note that the even-ness and odd-ness of the real and imaginary parts of the Fourier Transform of f depend essentially on the assumption that f is a real function). In particular, if f is a real function of t which is *even* in t, then we get

$$B(\omega) \equiv 0 \quad \text{and} \quad F(i\omega) = A(\omega) = 2 \int_{0}^{\infty} f(t) \cos \omega t \, dt .$$

That is, the Fourier Transform of f is wholly real.

On the other hand, if f is a real function of t which is *odd* in t, then

$$A(\omega) \equiv 0 \quad \text{and} \quad F(i\omega) = -iB(\omega) = -2i \int_0^\infty f(t) \sin \omega t \, dt$$

so that the Fourier transform of f is purely imaginary.

6.3.2 Suppose now that f is a (real) function which vanishes identically outside the finite interval $-T/2 < t < +T/2$ and which satisfies the Dirichlet conditions within that interval. Then

$$F(i\omega) = \int_{-\infty}^{+\infty} f(t)e^{-i\omega t}dt = \int_{-T/2}^{T/2} f(t)e^{-i\omega t}dt$$

and

$$f_T(t) \cong \frac{A_0}{T} + \frac{2}{T} \sum_{n=1}^\infty \{A_n \cos n\omega_0 t + B_n \sin n\omega_0 t\} = \frac{1}{T} \sum_{n=-\infty}^{+\infty} C_n e^{in\omega_0 t}$$

where

$$C_n = \int_{-T/2}^{T/2} f(t)\, e^{-in\omega_0 t}dt \equiv F(in\omega_0)$$

$$A_n = \int_{-T/2}^{T/2} f(t) \cos n\omega_0 t \, dt \equiv A(n\omega_0) \tag{6.14}$$

$$B_n = \int_{-T/2}^{T/2} f(t) \sin n\omega_0 t \, dt \equiv B(n\omega_0)$$

To illustrate this relationship between Fourier transforms and Fourier series consider the simple but important example afforded by the rectangular pulse shown in Fig. 6.3(a). The corresponding Fourier transform, shown in Fig. 6.3(b), is easily computed as follows:

$$P(i\omega) = \int_{-\infty}^{+\infty} p(t)e^{-i\omega t}dt = k \int_{-d/2}^{d/2} e^{-i\omega t}dt = \frac{2k}{\omega} \sin \omega d/2$$

$$= kd \left[\frac{\sin \omega d/2}{\omega d/2}\right]. \tag{6.15}$$

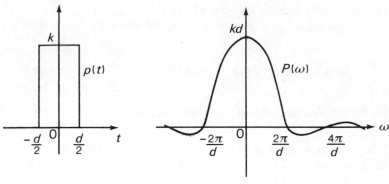

Fig. 6.3 (a).   Fig. 6.3 (b).

If T is any number greater than or equal to d then we can confirm the result of Question 3 of Exercises I by writing down at once the Fourier series for the function p_T obtained by periodically repeating p with period T (see Fig. 6.4(a)):

$$p_T(t) \equiv \sum_{n=-\infty}^{+\infty} p(t-nT) \cong \frac{1}{T} \sum_{n=-\infty}^{+\infty} P(in\omega_0) e^{in\omega_0 t}$$

$$= \frac{kd}{T} \sum_{n=-\infty}^{+\infty} \left[\frac{\sin n\omega_0 d/2}{n\omega_0 d/2}\right] e^{in\omega_0 t} = \frac{kd}{T}\left[1 + 2\sum_{n=1}^{\infty}\left[\frac{\sin n\omega_0 d/2}{n\omega_0 d/2}\right]\cos n\omega_0 t\right]$$

$$(6.16)$$

The Fourier coefficients C_n are given by the ordinates $P(in\omega_0)$ as shown in Fig. 6.4(b). With the usual interpretation of t as *time* and ω as (angular) *frequency* we might sum this up by saying that periodic **extension** of a function in the time domain corresponds to periodic **sampling** of the Fourier transform of the function in the frequency domain.

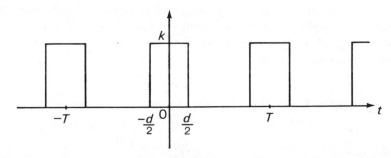

Fig. 6.4 (a).

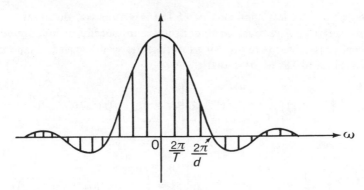

Fig. 6.4(b).

6.3.3 There is an **Inversion Theorem** for the Fourier Transform which is of crucial importance in what follows. A rigorous proof is beyond the scope of this text, and we give here only a heuristic derivation of the result. Nevertheless the argument does show how a judicious use of the delta function can make the result simple and conceptually clear.

We use the standard result (obtainable, for example, by elementary contour integration)

$$P \int_{-\infty}^{+\infty} \frac{e^{imx}}{x} \, dx = \pi i$$

where the symbol P denotes the Cauchy Principal Value of the integral and m is a positive constant. That is,

$$\int_{-\infty}^{+\infty} \frac{\sin mx}{x} \, dx = \pi \quad \text{and} \quad \int_{-\infty}^{+\infty} \frac{\cos mx}{x} \, dx = 0 \;.$$

Replacing m by $-m$ simply changes the sign of the first of these two real integrals and leaves the other unaltered. Hence if we replace m by the usual symbol t for the independent real variable we can write

$$\frac{1}{2\pi} \int_{-\infty}^{+\infty} \frac{e^{itx}}{ix} \, dx = \frac{1}{2} \, \text{sgn} \, t \equiv \begin{cases} +1/2 & \text{for } t > 0 \\ -1/2 & \text{for } t < 0 \end{cases} . \tag{6.17}$$

A formal differentiation of both sides of (6.17) with respect to t then yields

$$\frac{1}{2\pi} \int_{-\infty}^{+\infty} e^{itx} \, dx = \delta(t) \;. \tag{6.18}$$

The integral on the left-hand side of (6.18) is, of course, divergent, and it is clear that this equation must be understood symbolically or interpreted in an operational sense. That is to say, for all sufficiently well-behaved functions f, we should interpret (6.18) to mean that

$$\int_{-\infty}^{+\infty} f(t) \left[\frac{1}{2\pi} \int_{-\infty}^{+\infty} e^{i\omega t} \, d\omega \right] dt = \int_{-\infty}^{+\infty} f(t)\delta(t)\,dt = f(0)$$

or, more generally, that

$$\int_{-\infty}^{+\infty} f(\tau) \left[\frac{1}{2\pi} \int_{-\infty}^{+\infty} e^{i(t-\tau)\omega} \, d\omega \right] d\tau = \int_{-\infty}^{+\infty} f(\tau)\delta(t-\tau)\,d\tau = f(t)$$

$$(6.19)$$

With due disregard for any of the niceties involved in altering the order of the integration signs appearing in (6.19), we can rewrite this result in the form

$$f(t) = \frac{1}{2\pi} \int_{-\infty}^{+\infty} e^{i\omega t} \left[\int_{-\infty}^{+\infty} f(\tau) e^{-i\omega \tau} \, d\tau \right] d\omega = \frac{1}{2\pi} \int_{-\infty}^{+\infty} e^{i\omega t} F(i\omega)\,d\omega$$

$$(6.20)$$

where $F(i\omega)$ denotes, as usual, the Fourier Transform of f.

The validity of the inversion formula (6.20) obviously depends critically on the conditions which must be imposed on the function f to make the above assertions and manipulations justifiable. (This is of course the crux of any genuine proof of the result and is precisely what we have glossed over above in the phrase "sufficiently well-behaved".) For completeness we state here a form of the **Fourier Inversion Theorem** proper which is sufficiently comprehensive for most purposes and which can be established rigorously.

Let f be a (real or complex valued) function of a single real variable which

(a) is absolutely integrable over the interval $(-\infty, +\infty)$, and
(b) satisfies the Dirichlet conditions over every finite interval.

If $F(i\omega)$ denotes the Fourier Transform of f then at each point t we have

$$\frac{1}{2\pi} \int_{-\infty}^{+\infty} e^{i\omega t} F(i\omega)\,d\omega = \frac{1}{2} \{f(t+) + f(t-)\}. \qquad (6.21)$$

(At all points of continuity of f, (6.21) reduces to (6.20)).

6.3.4 As in the case of the Laplace Transform there are several important properties of the Fourier Transform which merit explicit mention. It is a simple matter to establish the following (see Exercises III, Question 2):

   (i) Linearity: $\mathcal{F}\{af_1(t) + bf_2(t)\} = aF_1(i\omega) + bF_2(i\omega)$
   (ii) $\mathcal{F}\{f(t-a)\} = \exp(-i\omega a)F(i\omega)$
   (iii) $\mathcal{F}\{f(t)\exp(-at)\} = F(i\omega + a)$
   (iv) $\mathcal{F}\{f'(t)\} = i\omega F(i\omega)$.

Moreover, there is a convolution theorem for the Fourier Transform which compares directly with that for the Laplace Transform:

Let

$$h(t) = f(t)*g(t) = \int_{-\infty}^{+\infty} f(t-\tau)g(\tau)d\tau$$

Then

$$H(i\omega) = \int_{-\infty}^{+\infty} e^{-i\omega t}dt \int_{-\infty}^{+\infty} f(t-\tau)g(\tau)d\tau$$

$$= \int_{-\infty}^{+\infty} g(\tau)d\tau \int_{-\infty}^{+\infty} f(t-\tau)e^{-i\omega t}dt = \int_{-\infty}^{+\infty} g(\tau)d\tau \int_{-\infty}^{+\infty} f(x)e^{-i\omega(x+\tau)}dx$$

$$= \int_{-\infty}^{+\infty} e^{-i\omega \tau}g(\tau)d\tau \int_{-\infty}^{+\infty} e^{-i\omega x}f(x)dx = F(i\omega)G(i\omega) .$$

Using the Fourier Inversion Theorem we can derive a dual result which applies to the convolution of Fourier Transforms themselves. Thus, given that $F(i\omega) \equiv \mathcal{F}\{f(t)\}$ and that $G(i\omega) \equiv \mathcal{F}\{g(t)\}$ are themselves both absolutely integrable functions over $(-\infty, +\infty)$, consider their convolution

$$H(i\omega) = \int_{-\infty}^{+\infty} F\{i(\omega - v)\}G(iv)dv .$$

From the Inversion Theorem,

$$\mathcal{F}^{-1}\{H(i\omega)\} \equiv h(t) \cong \frac{1}{2\pi} \int_{-\infty}^{+\infty} H(i\omega)e^{i\omega t}d\omega$$

$$= \frac{1}{2\pi} \int_{-\infty}^{+\infty} e^{i\omega t}d\omega \int_{-\infty}^{+\infty} F\{i(\omega - v)\}G(iv)$$

$$= \frac{1}{2\pi} \int_{-\infty}^{+\infty} G(iv)dv \int_{-\infty}^{+\infty} e^{i\omega t}F\{i(\omega - v)\} dv$$

$$= \frac{1}{2\pi} \int_{-\infty}^{+\infty} G(iv)\,dv \int_{-\infty}^{+\infty} F(i\mu)e^{i(\mu+v)t}\,d\mu$$

$$= \frac{1}{2\pi} \int_{-\infty}^{+\infty} G(iv)e^{ivt}\,dv \int_{-\infty}^{+\infty} F(i\mu)e^{i\mu t}\,d\mu = g(t)\{2\pi f(t)\}\,.$$

Thus we have the dual results

$$\mathcal{F}\{f(t) * g(t)\} = F(i\omega)G(i\omega)$$

$$\mathcal{F}\{f(t)g(t)\} = \frac{1}{2\pi}[F(i\omega) * G(i\omega)]\,. \quad\left.\right\} \quad (6.22)$$

Exercises III

1. If $\mathcal{F}\{f(t)\} = F(i\omega)$ show that

(a) $\mathcal{F}\{f(t-a)\} = \exp(-i\omega a)F(i\omega)$,

(b) $\mathcal{F}\{f(t)\exp(-at)\} = F(i\omega + a)$,

(c) $\mathcal{F}\{f(at)\} = \frac{1}{a}F(i\omega/a)$,

(d) $\mathcal{F}\{f'(t)\} = i\omega F(i\omega)$, where in (d) we assume that the classical derivative of f exists and is an absolutely integrable function over $(-\infty, +\infty)$.

2. If $h_1(t) = \begin{cases} 1 & \text{for } |t| < a/2 \\ 0 & \text{otherwise} \end{cases}$, and $h_2(t) = \begin{cases} 1 & \text{for } |t| < b/2 \\ 0 & \text{otherwise} \end{cases}$

find the function $h_3(t) = h_1(t) * h_2(t)$ and confirm by direct calculation that

$$H_3(i\omega) = H_1(i\omega)H_2(i\omega)\,.$$

3. Find the Fourier Transforms of each of the following functions:

(a) $\exp(-|t|)$; (b) $\operatorname{sgn} t \exp(-|t|)$; (c) $\exp(-t^2/2)$;

(d) $f_1(t) = \begin{cases} 1 - t^2 & \text{for } |t| < 1 \\ 0 & \text{otherwise} \end{cases}$; (e) $f_2(t) = \begin{cases} \sin t & \text{for } 0 < t < \pi \\ 0 & \text{otherwise} \end{cases}$;

(f) $f_3(t) = \begin{cases} \cos t & \text{for } -\pi/2 < t < +\pi/2 \\ 0 & \text{otherwise} \end{cases}$

[Hint: in (c) use the fact that $\displaystyle\int_{-\infty}^{+\infty} e^{-x^2}\,dx = \sqrt{\pi}$ and recall that

$$t^2 + 2at = (t + a)^2 - a^2].$$

4. Let f be a real function of t which is absolutely integrable over $(-\infty, +\infty)$ and continuous everywhere. Use the Fourier Inversion Theorem to show that

$$f(t) = \frac{1}{\pi} \int_0^{+\infty} d\omega \int_{-\infty}^{+\infty} f(\tau) \cos \omega \, (t - \tau) d\tau \, .$$

Deduce that, if $f(t)$ is even, then

$$f(t) = \frac{2}{\pi} \int_0^{+\infty} \cos \omega \, t \, d\omega \int_0^{+\infty} f(\tau) \cos \omega\tau \, d\tau$$

while, if f is odd, then

$$f(t) = \frac{2}{\pi} \int_0^{+\infty} \sin \omega t \, d\omega \int_0^{+\infty} f(\tau) \sin \omega\tau \, d\tau \, .$$

5. By using the results of (3) and (4) above evaluate the integrals

(a) $\displaystyle\int_0^{+\infty} \frac{\cos xt}{x^2 + 1} \, dx \, ;$ (b) $\displaystyle\int_0^{+\infty} \left[\frac{\omega \cos \omega - \sin \omega}{\omega^3} \right] d\omega \, .$

6.4 GENERALISED FOURIER TRANSFORMS

6.4.1 A function $n(t)$ is said to be a **null-function** on the interval $-\infty < t < +\infty$ if

$$\int_{-\infty}^{+\infty} |n(t)| \, dt = 0 \, .$$

Two functions which differ only by a null function are said to be equal **almost everywhere**, and we write $f_1(t) \underset{\text{a.e.}}{=} f_2(t)$. This means, in effect, that the set of points at which the two functions differ in value is a negligibly small set in the sense that it is of zero total length. We have already used this idea in connection with Laplace Transform inversion (Sec. 5.1.4), and the point will be taken up again in Chapter 8. In the present context we can use it to give an alternative statement of the Fourier Inversion Theorem.

Let f be absolutely integrable over $(-\infty, +\infty)$. The Fourier Transform of f is a bounded, continuous, function given for all ω by

$$F(i\omega) = \int_{-\infty}^{+\infty} f(t) e^{-i\omega t} dt \tag{6.23}$$

and we have

$$f(t) \underset{\text{a.e.}}{=} \lim_{R \to \infty} \frac{1}{2\pi} \int_{-R}^{+R} F(i\omega) e^{i\omega t} d\omega \, . \tag{6.24}$$

In this statement we emphasise first that the inversion integral in (6.24) is not generally an absolutely convergent integral (in contrast to the direct Fourier integral in (6.23)). Secondly we know that at points of discontinuity of f the right-hand side of (6.24) may fail to converge to the values assumed by the function; the use of the symbol "$\underset{\text{a.e.}}{=}$" indicates that the totality of such points is negligibly small.

Now for many purposes we need to work with functions which, while not necessarily absolutely integrable over $(-\infty, +\infty)$, are **square-integrable** over that range; that is to say, we have

$$\int_{-\infty}^{+\infty} |f(t)|^2 \, \mathrm{d}t < +\infty .$$

In such a case the Fourier Transform of f can be defined, at least almost everywhere, in terms of an integral which is now not necessarily absolutely convergent:

$$F(i\omega) \underset{\text{a.e.}}{=} \lim_{R \to \infty} \int_{-R}^{+R} f(t)\,e^{-i\omega t}\,\mathrm{d}t . \qquad (6.25)$$

The function F may not be continuous everywhere, nor even bounded. However, it turns out that it is certainly a square-integrable function itself over the range $-\infty < \omega < +\infty$, and that the inversion formula still holds in the form (6.24).

For example, the function $f(t) = \dfrac{1}{\pi t} \sin \pi t$ is not absolutely integrable over $(-\infty, +\infty)$ but we do have $\displaystyle\int_{-\infty}^{+\infty} |f(t)|^2 \, \mathrm{d}t = 1$.

The Fourier Transform $F(i\omega)$ is given by

$$F(i\omega) = \begin{cases} 1 & \text{for } |\omega| < \pi \\ 0 & \text{for } |\omega| > \pi \end{cases}$$

and is plainly discontinuous at $\omega = \pm\pi$.

In spite of this extension to functions of finite energy, there still remains a variety of important and useful functions which are neither absolutely integrable nor square-integrable, and it is desirable to extend the meaning of Fourier Transform yet again so as to apply to these. It is this which requires us once more to introduce generalised functions.

6.4.2 Applying the sampling property of the delta function to the Fourier integral itself we get

$$\int_{-\infty}^{+\infty} e^{-i\omega t}\delta(t)\,\mathrm{d}t = 1 \qquad (6.26)$$

or, more generally,

$$\int_{-\infty}^{+\infty} e^{-i\omega t}\delta(t-a)\,dt = e^{-i\omega a} . \qquad (6.27)$$

This suggests that we define the Fourier Transform of $\delta(t-a)$ to be $\exp(-i\omega a)$; the inversion formula would then be formally satisfied since, using (6.18), we get

$$\frac{1}{2\pi}\int_{-\infty}^{+\infty} e^{i\omega t}e^{-i\omega a}\,d\omega = \frac{1}{2\pi}\int_{-\infty}^{+\infty} e^{i\omega(t-a)}\,d\omega = \delta(t-a) .$$

Dually we can apply the sampling property of the delta function to the Fourier inversion integral:

$$\frac{1}{2\pi}\int_{-\infty}^{+\infty} e^{i\omega t}\delta(\omega-\alpha)\,d\omega = \frac{1}{2\pi}e^{i\alpha t}$$

and similarly

$$\frac{1}{2\pi}\int_{-\infty}^{+\infty} e^{i\omega t}\delta(\omega+\alpha)\,d\omega = \frac{1}{2\pi}e^{-i\alpha t} .$$

Thus, recalling that the Fourier Transform is defined in general for complex valued functions, these results suggest that we should have

$$\mathcal{F}\{e^{i\alpha t}\} = 2\pi\delta(\omega-\alpha)$$
$$\mathcal{F}\{e^{-i\alpha t}\} = 2\pi\delta(\omega+\alpha) . \qquad (6.28)$$

So far as real functions of t are concerned the equations (6.28) yield the following results:

$$\mathcal{F}\{\cos \alpha t\} = \pi\{\delta(\omega-\alpha)+\delta(\omega+\alpha)\}$$
$$\mathcal{F}\{\sin \alpha t\} = \frac{\pi}{i}\{\delta(\omega-\alpha)-\delta(\omega+\alpha)\} . \qquad (6.29)$$

In particular, taking $\alpha = 0$, we find that the generalised Fourier Transform of the constant function $f(t) \equiv 1$ is simply $2\pi\,\delta(\omega)$. This in turn allows us to offer a definition of the Fourier Transform of the unit step function. For we have

$$u(t) = \frac{1}{2} + \frac{1}{2}\operatorname{sgn} t$$

and from (6.17) we know that $2/i\omega$ is a suitable choice for the Fourier Transform of the function sgn t in the sense that

$$\operatorname{sgn} t = \frac{1}{2\pi}\int_{-\infty}^{+\infty} \frac{2}{i\omega}e^{i\omega t}\,d\omega .$$

Accordingly it follows that we should set

$$\mathcal{F}\{u(t)\} = \pi\delta(\omega) + \frac{1}{i\omega}. \tag{6.30}$$

Finally, recall that we have already established some formal "Fourier series" expansions for periodic trains of delta functions:

$$\sum_{n=-\infty}^{+\infty} \delta(t-nT) = \frac{1}{T}\left[1 + 2\sum_{n=1}^{\infty}\cos n\omega_0 t\right] = \frac{1}{T}\sum_{n=-\infty}^{+\infty} e^{in\omega_0 t} \tag{6.31}$$

We now examine the possibility of extending the definition of Fourier Transform to apply to these symbolic expressions. First, since the Fourier Transform of $\delta(t-nT)$ is $e^{-inT\omega}$, we ought to get

$$\mathcal{F}\left[\sum_{n=-\infty}^{+\infty}\delta(t-nT)\right] = \sum_{n=-\infty}^{+\infty} e^{-inT\omega} \equiv \sum_{n=-\infty}^{+\infty} e^{inT\omega}. \tag{6.32}$$

On the other hand, the Fourier Transform of $e^{in\omega_0 t}$ has been seen to be $2\pi\delta(\omega - n\omega_0)$, and that of $\cos n\omega_0 t$ to be $\pi[\delta(\omega - n\omega_0) + \delta(\omega + n\omega_0)]$. Accordingly the Fourier Transform of the exponential and trigonometric expressions appearing in (6.31) would appear to be

$$\mathcal{F}\left[\frac{1}{T}\left[1 + 2\sum_{n=1}^{\infty}\cos n\omega_0 t\right]\right] = \mathcal{F}\left[\frac{1}{T}\sum_{n=-\infty}^{+\infty} e^{in\omega_0 t}\right] = \omega_0\sum_{n=-\infty}^{+\infty}\delta(\omega-n\omega_0). \tag{6.33}$$

Comparing (6.32) and (6.33),

$$\sum_{n=-\infty}^{+\infty} e^{in\omega T} = \frac{2\pi}{T}\sum_{n=-\infty}^{+\infty}\delta\left(\omega - \frac{n2\pi}{T}\right) \tag{6.34}$$

where, for the sake of clarity, we replace ω_0 by $2\pi/T$. Note that once again the relation between the left-hand and right-hand sides of (6.34) is formally that of a Fourier series to the periodic function which it represents.

6.4.3 We shall now apply the considerations of the preceding section to derive the generalised Fourier Transform of an arbitrary periodic function. Let f be a bounded, continuous, function of t which vanishes outside the finite interval $(-T/2, +T/2)$ and which satisfies the Dirichlet conditions within that interval. Using the fact that the Fourier Transform of $f(t-nT)$ is $e^{-inT\omega}F(i\omega)$, we get

$$\mathcal{F}\left[\sum_{n=-\infty}^{+\infty} f(t-nT)\right] = \sum_{n=-\infty}^{+\infty} F(i\omega)e^{-inT\omega} = F(i\omega)\sum_{n=-\infty}^{+\infty} e^{-inT\omega}$$

$$= \omega_0 \sum_{n=-\infty}^{+\infty} F(i\omega)\delta(\omega - n\omega_0) = \frac{2\pi}{T} \sum_{n=-\infty}^{+\infty} F\left(\frac{i2\pi n}{T}\right) \delta\left(\omega - \frac{2\pi n}{T}\right).$$

Now formally apply the inversion formula to this last expression,

$$\sum_{n=-\infty}^{+\infty} f(t-nT) \cong \frac{1}{2\pi} \int_{-\infty}^{+\infty} e^{i\omega t} \left[\frac{2\pi}{T} \sum_{n=-\infty}^{+\infty} \left(F \frac{i2\pi n}{T}\right) \delta\left(\omega - \frac{2\pi n}{T}\right)\right] d\omega$$

$$= \frac{1}{T} \sum_{n=-\infty}^{+\infty} \left(F \frac{i2\pi n}{T}\right) e^{in2\pi t/T}$$

and we recover the usual Fourier series expansion of the periodic extension of f.

This result does support the alleged formal equivalence of the expressions in (6.31). For we know that

$$\sum_{n=-\infty}^{+\infty} f(t-nT) = f(t) * \sum_{n=-\infty}^{+\infty} \delta(t-nT)$$

and so we ought to expect that the Fourier Transform of the right-hand side should be the product

$$F(i\omega) \times \mathcal{F}\left[\sum_{n=-\infty}^{+\infty} \delta(t-nT)\right].$$

That this is the case follows precisely because of the derivation of the Fourier Transform of the impulse train from the "Fourier series" expansions of (6.31).

Exercises IV

1. Find the (generalised) Fourier Transforms of each of the following:

    (a) $f_1(t) = (1 - e^{-at})u(t)$; (b) $f_2(t) = \cos^2 at$; (c) $f_3(t) = \sin^2 at$.

2. Delta functions are located at the points

    $$t = 0, T, 2T, \ldots, (2N-1)T.$$

    Calculate the Fourier Transform of this train of delta functions and hence find the inverse Fourier Transform of

    $$\frac{\sin N\omega T}{\sin \omega T/2}.$$

3. A real-valued, continuous, function f, with Fourier Transform $F(i\omega)$, is periodically sampled at the points $t = nT$ (where $n = 0, \pm 1, \pm 2, \ldots$). This

may be regarded as an operation which transforms $f(t)$ into the generalised function

$$f^*(t) \equiv \sum_{n=-\infty}^{+\infty} f(nT)\delta(t-nT)$$

by means of the formal multiplication

$$f^*(t) \equiv \sum_{n=-\infty}^{+\infty} f(nT)\delta(t-nT) = f(t) \times \sum_{n=-\infty}^{+\infty} \delta(t-nT) \, .$$

Assuming that the infinite series concerned do converge, show formally that

$$\mathcal{F}\text{-}[f^*(t)] = \frac{1}{T} \sum_{n=-\infty}^{+\infty} F\left[i\left(\omega - \frac{2\pi n}{T}\right)\right] = \sum_{n=-\infty}^{+\infty} f(nT)\,e^{-in\omega T} \, .$$

4. Suppose that in Question 3 the Fourier Transform $F(i\omega)$ of the function f vanishes identically outside the interval

$$\left(-\frac{\pi}{T}, +\frac{\pi}{T}\right).$$

Let $h(t)$ be the impulse response function of a time-invariant linear system whose transfer function, $H(i\omega)$, is equal to 1 for all ω such that $|\omega| < \pi/T$ and is equal to 0 otherwise. (Such a T.I.L.S. is often called an *ideal, low-pass, filter*.) Find the output signal which results when the sampled function $f^*(t)$ is applied as an input, and hence show that

$$f(t) = \sum_{n=-\infty}^{+\infty} f(nT) \cdot \left[\frac{\sin(t-nT)\pi/T}{(t-nT)\pi/T}\right].$$

CHAPTER 7

Other Types of Generalised Function

7.1 THE FINITE PART OF A DIVERGENT INTEGRAL

7.1.1 In the preceding chapters we have been concerned with the development of manipulative rules proper to certain mathematical entities which we have called generalised functions. These were introduced in the first place to allow the processes of elementary calculus to be extended to functions with jump discontinuities. In each case it turned out that the generalised function concerned was defined not in the usual pointwise sense appropriate to ordinary functions but rather in terms of some characteristic operation acting on (suitably restricted classes of) ordinary functions. It is natural to ask if a similar approach could be made to the problem of defining derivatives for functions with unbounded discontinuities.

To begin with we recall the elementary formula for integration by parts. If f and g are continuously differentiable everywhere and if $[a,b]$ is any finite, closed, interval then

$$\int_a^b f(t)g'(t)\mathrm{d}t = -\int_a^b f'(t)g(t)\mathrm{d}t + f(b)g(b) - f(a)g(a) . \qquad (7.1)$$

In particular if f happens to vanish identically outside $[a,b]$ then (7.1) becomes

$$\int_{-\infty}^{+\infty} f(t)g'(t)\mathrm{d}t = [f(t)g(t)]_a^b - \int_a^b f'(t)g(t)\mathrm{d}t$$

$$= \int_{-\infty}^{+\infty} \{-f'(t)\}\, g(t)\mathrm{d}t . \qquad (7.2)$$

Equation (7.2) remains true if we relax the conditions on g and stipulate merely that it be *absolutely continuous* on $[a,b]$. (See Problem 3 of Exercises I of

Chapter 2.) In this case g need not be differentiable everywhere. As an example we could take the function

$$g(t) = u(t)\sqrt{t} \ .$$

This is an absolutely continuous function which is differentiable everywhere except at the origin; its derivative is $u(t)/2\sqrt{t}$ and this is actually unbounded on any neighbourhood of the origin. Nevertheless, as is easily confirmed, equation (7.2) holds for any continuously differentiable function f which vanishes identically outside some finite interval. (The integral on the left-hand side of (7.2) always exists because $g'(t)$ is a *locally integrable* function.)

7.1.2 If we take $g(t) = u(t)$ then only the integral on the right-hand side of equation (7.2) exists in the classical sense, and we have

$$\int_{-\infty}^{+\infty} \{-f'(t)\}\,g(t)\,\mathrm{d}t = \int_0^b \{-f'(t)\}\,\mathrm{d}t = [-f(t)]_0^b = f(0) \ .$$

This value is, of course, precisely what we would wish to attribute to the (symbolic) integral which now does duty for the left-hand side of equation (7.2). In fact we could equally well base the definition of the delta function itself on the equation.

$$\int_{-\infty}^{+\infty} f(t)\delta(t)\,\mathrm{d}t = \int_{-\infty}^{+\infty} \{-f'(t)\}\,u(t)\,\mathrm{d}t \qquad\qquad (7.3)$$

in the sense that the meaning of the left-hand side is to be understood simply as the value which the (well-defined) right-hand side assigns to it. The only difference from what has gone before is that the operation of the delta function is limited by (7.3) to a more restricted class of functions f than we have found to be desirable.

7.1.3 Now let us use a similar technique to try to extend the concept of derivative so that it applies usefully and meaningfully to functions with infinite discontinuities. As a first example we shall take the function

$$g(t) = -2u(t)/\sqrt{t}.$$

This is differentiable in the classical sense everywhere except at the origin, and its classical derivative is given by

$$g'(t) = u(t)/t^{3/2} \ .$$

The function $g'(t)$ is unbounded in any neighbourhood of the origin, and in this respect the situation appears at first to be similar to that of the absolutely continuous function $u(t)\sqrt{t}$, and its derivative $u(t)/2\sqrt{t}$, already discussed in

Sec. 7.1.1. However, this time the derivative $g'(t)$ is *not* a locally integrable function; it is not integrable over any finite interval containing the origin. Hence if we write

$$\int_{-\infty}^{+\infty} f(t)g'(t)\,dt = \int_{-\infty}^{+\infty} \{-f'(t)\}\{-2u(t)/\sqrt{t}\}\,dt \tag{7.4}$$

then this will not be a valid equation for all continuously differentiable functions f which vanish outside a finite interval; the right-hand side always exists, but the left-hand side generally does not. More precisely, for the right-hand side we have

$$\int_{-\infty}^{+\infty} \{-f'(t)\}\{-2u(t)/\sqrt{t}\}\,dt = 2\int_0^b \frac{f'(t)}{\sqrt{t}}\,dt \tag{7.5}$$

for some finite number b, and this is an absolutely convergent integral. On the other hand we can analyse the left-hand side of (7.4) as follows:

$$\int_{-\infty}^{+\infty} f(t)\{u(t)t^{-3/2}\}\,dt = \lim_{\epsilon \downarrow 0} \int_\epsilon^\infty f(t)t^{-3/2}\,dt$$

$$= -2\lim_{\epsilon \downarrow 0} \int_\epsilon^b f(t)\frac{d}{dt}\left(\frac{1}{\sqrt{t}}\right)dt = -2\lim_{\epsilon \downarrow 0}\left[\frac{-f(\epsilon)}{\sqrt{\epsilon}} - \int_\epsilon^b \frac{f'(t)}{\sqrt{t}}\,dt\right] \tag{7.6}$$

making an entirely legitimate use of integration by parts, and of the fact that $f(b) = 0$.

   Now, since f is continuously differentiable, we can appeal to the First Mean Value Theorem of the differential calculus (that is, Taylor's Theorem with $n = 1$) and write

$$f(t) = f(0) + tf'(\theta t) \tag{7.7}$$

where θ is some number lying between 0 and 1. In particular this gives

$$\frac{f(\epsilon)}{\sqrt{\epsilon}} = \frac{f(0)}{\sqrt{\epsilon}} + \sqrt{\epsilon}\,f'(\theta\epsilon).$$

It follows that $f(\epsilon)/\sqrt{\epsilon}$ tends to 0 with ϵ if and only if $f(0) = 0$; if $f(0) \neq 0$ then this ratio tends either to $+\infty$ or to $-\infty$. Hence if $f(0) = 0$ then the limit of (7.6) exists and is equal to the finite value given by (7.5); in this case the equation (7.4) is a valid one. If, on the other hand, $f(0) \neq 0$ then the form of (7.6) allows us to separate out the required "finite part" of the divergent integral which now appears as the left-hand side of (7.4). To do this we have only to remove the term $2f(\epsilon)/\sqrt{\epsilon}$ which, as the above argument shows, is what causes the limit in (7.6) to become infinite. The remaining well-defined quantity is called the **Hadamard Finite Part**[†] of the divergent integral concerned, and we write

$$\text{Fp}\int_{-\infty}^{+\infty} f(t)\{u(t)t^{-3/2}\}\,dt \equiv \lim_{\epsilon\downarrow 0}\left[\int_{\epsilon}^{\infty} f(t)t^{-3/2}\,dt - 2f(\epsilon)/\sqrt{\epsilon}\right] \quad (7.8)$$

the symbol "Fp" expressing the fact that we are extracting a certain finite number from an integral which is actually divergent.

7.1.4 The preceding analysis may be summed up as follows. We require to define an "operational" derivative, Dg, of the function $g(t) = -2u(t)/\sqrt{t}$. This must satisfy the equation

$$\int_{-\infty}^{+\infty} f(t)\{\text{D}g(t)\}\,dt = \int_{-\infty}^{+\infty}\{-f'(t)\}\,g(t)\,dt$$

at least for all continuously differentiable functions f which vanish outside a finite interval. We cannot identify Dg with the ordinary derivative g' since this generally makes the integral on the left of this equation divergent. Instead we agree to regard Dg as a new type of generalised function. It is usual to refer to Dg as a **pseudo-function**, and to denote it by the symbol Pf $\{u(t)t^{-3/2}\}$. Just as the generalised function δ is characterised by the operation which carries each continuous function f into the corresponding number $f(0)$, so this pseudo-function is characterised by the operation which carries each continuously differentiable function f, vanishing outside some finite interval, into the number given by the limit in (7.8).

We can express the finite part of the integral in (7.8) in a more convenient form by using again the Taylor expansion of (7.7). We have

$$\int_{-\infty}^{+\infty} f(t)\{u(t)t^{-3/2}\}\,dt = \lim_{\epsilon\downarrow 0}\int_{\epsilon}^{\infty} f(t)t^{-3/2}\,dt$$

$$= \lim_{\epsilon\downarrow 0}\left[\int_{\epsilon}^{\infty}\frac{f(0)}{t^{3/2}}\,dt + \int_{\epsilon}^{\infty}\frac{f'(\theta t)}{\sqrt{t}}\,dt\right] = \lim_{\epsilon\downarrow 0}\left[\frac{2f(0)}{\sqrt{\epsilon}} + \int_{\epsilon}^{\infty}\frac{f'(\theta t)}{\sqrt{t}}\,dt\right]. \quad (7.9)$$

† J. Hadamard, 'Lectures on Cauchy's Problem', Yale University Press, 1923.

Comparing (7.9) with (7.8) it is clear that we can use any one of the following formulations:

$$\int_{-\infty}^{+\infty} f(t)\,\text{Pf}\,\{u(t)/t^{3/2}\}\cdot dt \equiv \text{Fp} \int_{-\infty}^{+\infty} f(t)\,\{u(t)/t^{3/2}\}\,dt$$

$$= \int_0^\infty \{f'(\theta t)/\sqrt{t}\}\,dt \tag{7.10}$$

$$= \int_0^\infty \frac{f(t) - f(0)}{t^{3/2}}\,dt . \tag{7.11}$$

7.2 PSEUDO-FUNCTIONS

7.2.1 The analysis leading to the definition of the pseudo-function $\text{Pf}\,\{u(t)t^{-3/2}\}$ can be immediately generalised. Let $h(t) = u(t)/t^{\alpha+1}$ where $0 < \alpha < 1$. Take an arbitrary continuously differentiable function f, vanishing outside some finite interval, say $[a,b]$. Then

$$\int_\epsilon^\infty \frac{f(t)}{t^{\alpha+1}}\,dt = \int_\epsilon^\infty \left[\frac{f(0)}{t^{\alpha+1}} + \frac{f'(\theta t)}{t^\alpha} \right] dt$$

$$= \int_\epsilon^\infty \frac{f'(\theta t)}{t^\alpha}\,dt + f(0)\left[\frac{t^{-\alpha}}{-\alpha} \right]_\epsilon^\infty = \int_\epsilon^\infty \frac{f'(\theta t)}{t^\alpha}\,dt + \frac{f(0)}{\alpha\epsilon^\alpha}$$

where we make use of the Taylor expansion

$$f(t) = f(0) + tf'(\theta t), \quad 0 < \theta < 1 .$$

The last integral converges absolutely as ϵ tends to zero, and the second term tends to plus or minus infinity unless $f(0) = 0$. Hence

$$\text{Fp} \int_{-\infty}^{+\infty} f(t)h(t)\,dt = \lim_{\epsilon \downarrow 0} \int_\epsilon^\infty \frac{f'(\theta t)}{t^\alpha}\,dt = \lim_{\epsilon \downarrow 0} \int_\epsilon^\infty \left[\frac{f(t) - f(0)}{t^{\alpha+1}} \right] dt . \tag{7.12}$$

This (absolutely convergent) integral defines the pseudo-function $\text{Pf}\,\{u(t)/t^{\alpha+1}\}$. This is the generalised derivative of the ordinary function $-u(t)/\alpha t^\alpha$, for we have

$$\int_\epsilon^\infty \left[\frac{f(t) - f(0)}{t^{\alpha+1}} \right] dt = \left[\left(\frac{t^{-\alpha}}{-\alpha} \right) \left[f(t) - f(0) \right] \right]_\epsilon^\infty + \frac{1}{\alpha} \int_\epsilon^\infty f'(t) t^{-\alpha}\,dt$$

$$= \int_\epsilon^\infty \{-f'(t)\} \left[\frac{t^{-\alpha}}{-\alpha} \right] dt + \frac{1}{\alpha} \left[\frac{f(\epsilon) - f(0)}{\epsilon^\alpha} \right]$$

where the last term tends to 0 with ϵ since $f(\epsilon) - f(0) = \epsilon f'(\theta \epsilon)$ and $0 < \alpha < 1$. As a result we may write

$$\int_{-\infty}^{+\infty} f(t) \text{Pf} \left[\frac{u(t)}{t^{\alpha+1}} \right] dt \equiv \text{Fp} \int_{-\infty}^{\infty} \left[f(t) \frac{u(t)}{t^{\alpha+1}} \right] dt \ .$$

$$= \int_{-\infty}^{\infty} \{-f'(t)\} \left[\frac{-u(t)}{\alpha t^{\alpha}} \right] dt \ . \tag{7.13}$$

7.2.2 Now consider the function $k(t) = u(t)/t^{\alpha+2}$, where $0 < \alpha < 1$. As before we resort to a Taylor expansion in order to separate out the required finite part of a normally divergent integral. For an arbitrary twice-continuously differentiable function (again assumed to vanish outside some finite interval) we have

$$f(t) = f(0) + tf'(t) + \frac{t^2}{2} f''(\theta t)$$

where θ lies between 0 and 1. Then we get

$$\int_{\epsilon}^{\infty} \frac{f(t)}{t^{\alpha+2}} dt = \int_{\epsilon}^{\infty} \left[\frac{f(0)}{t^{\alpha+2}} + \frac{f'(0)}{t^{\alpha+1}} + \frac{f''(\theta t)}{2 t^{\alpha}} \right] dt$$

$$= \int_{\epsilon}^{\infty} \frac{f''(\theta t)}{2 t^{\alpha}} dt + \frac{f'(0)}{\alpha \epsilon^{\alpha}} + \frac{f(0)}{(\alpha+1)\epsilon^{\alpha+1}}$$

which diverges as ϵ tends to 0 unless $f'(0) = f(0) = 0$. The Hadamard finite part therefore takes the form

$$\text{Fp} \int_{-\infty}^{+\infty} f(t) k(t) dt = \lim_{\epsilon \downarrow 0} \int_{\epsilon}^{\infty} \frac{f''(\theta t)}{2 t^{\alpha}} dt$$

$$= \int_{0}^{\infty} \left[\frac{f(t) - f(0) - tf'(0)}{t^{\alpha+2}} \right] dt \ . \tag{7.14}$$

The mapping which carries each given f into the number specified by (7.14) defines the pseudo-function Pf $\{u(t)/t^{\alpha+2}\}$. Also, integrating by parts gives

the value of the integral $\quad \int_{\epsilon}^{\infty} \left[\frac{f(t) - f(0) - tf'(0)}{t^{\alpha+2}} \right] dt \quad$ as

$$\left[\frac{t^{-(\alpha+1)}}{-(\alpha+1)}\left\{f(t)-f(0)-tf'(0)\right\}\right]_\epsilon^\infty + \frac{1}{\alpha+1}\int_\epsilon^\infty \left[\frac{f'(t)-f'(0)}{t^{\alpha+1}}\right]dt$$

$$=\frac{1}{\alpha+1}\int_\epsilon^\infty\left[\frac{f'(t)-f'(0)}{t^{\alpha+1}}\right]dt+\frac{1}{\alpha+1}\left[\frac{f(\epsilon)-f(0)-\epsilon f'(0)}{\epsilon^{\alpha+1}}\right].$$

In the limit as ϵ tends to 0 the last integral is seen, from (7.12), to converge to the limit

$$\int_0^\infty\left[\frac{f'(t)-f'(0)}{(\alpha+1)t^{\alpha+1}}\right]dt=\int_{-\infty}^{+\infty}\left\{-f'(t)+f'(0)\right\}\left[\frac{-u(t)}{(\alpha+1)t^{\alpha+1}}\right]dt$$

$$\equiv\int_{-\infty}^\infty\left\{-f'(t)\right\}\mathrm{Pf}\left[\frac{-u(t)}{(\alpha+1)t^{\alpha+1}}\right]dt.$$

The remaining terms tend to zero with ϵ since

$$f(\epsilon)-f(0)-\epsilon f'(0)=\frac{\epsilon^2}{2}f''(\theta\epsilon)$$

and $0<\alpha<1$. Thus $\mathrm{Pf}\left\{u(t)/t^{\alpha+2}\right\}$ behaves as the generalised (that is, operational) derivative of the pseudo-function $\mathrm{Pf}\left\{-u(t)/(\alpha+1)t^{\alpha+1}\right\}$ in the sense that

$$\int_{-\infty}^{+\infty}f(t)\mathrm{Pf}\left[\frac{u(t)}{t^{\alpha+2}}\right]dt=\int_{-\infty}^{+\infty}\left\{-f'(t)\right\}\mathrm{Pf}\left[\frac{-u(t)}{(\alpha+1)t^{\alpha+1}}\right]dt$$

at least for every twice-continuously differentiable function f which vanishes outside some finite interval. Note that it is now generally necessary to compute Hadamard finite parts of divergent integrals in order to obtain *both* sides of the "integration by parts" equation.

7.2.3 Clearly we could go on to develop similar pseudo-functions of the form $\mathrm{Pf}\left\{u(t)/t^{\alpha+n}\right\}$, where $0<\alpha<1$ and n is integral. However, it is time to look more closely at those cases which we seem to be deliberately omitting, namely the pseudo-functions corresponding to the ordinary functions of the form $u(t)/t^m$ where m is a positive integer. In fact the same general principles continue to apply, but there are points of detail which demand special attention. Consider first the function $u(t)/t$. If we write

$$\int_\epsilon^\infty \frac{f(t)}{t}\,dt = \int_1^\infty \frac{f(t)}{t}\,dt + \int_\epsilon^1 \left[\frac{f(0)}{t} + f'(\theta t)\right] dt$$

$$= \int_1^\infty \frac{f(t)}{t}\,dt + \int_\epsilon^1 f'(\theta t)\,dt - f(0)\log\epsilon\ ,$$

then the finite part is unequivocally defined as

$$\text{Fp}\int_{-\infty}^{+\infty} f(t)\frac{u(t)}{t}\,dt = \lim_{\epsilon\downarrow 0}\left[\int_1^\infty \frac{f(t)}{t}\,dt + \int_\epsilon^1 \left[\frac{f(t)-f(0)}{t}\right]dt\right] \qquad (7.15)$$

An integration by parts gives

$$\text{Fp}\int_{-\infty}^{+\infty} f(t)\frac{u(t)}{t}\,dt = \lim_{\epsilon\downarrow 0}\left[\int_\epsilon^\infty \{-f'(t)\}\log t\ dt - \{f(\epsilon)-f(0)\}\log\epsilon\right]$$

and we have only to note that

$$f(\epsilon) - f(0) = \epsilon f'(\theta\epsilon)$$

where $|f'(\theta\epsilon)|$ is bounded and $\lim_{\epsilon\downarrow 0} (\epsilon\log\epsilon) = 0$. Thus the locally-integrable function $u(t)\log|t|$ has for its derivative the pseudo-function $\text{Pf}\{u(t)/t\}$:

$$\int_{-\infty}^{+\infty} f(t)\text{Pf}\left[\frac{u(t)}{t}\right]dt \equiv \text{Fp}\int_{-\infty}^{+\infty} f(t)\frac{u(t)}{t}\,dt = \int_{-\infty}^{+\infty} \{-f'(t)\}\,u(t)\log|t|\,dt\ .$$

For $m = 2, 3, \ldots$, the position is complicated by the somewhat unexpected appearance of derivatives of delta functions. As an illustration we shall look briefly at the case $m = 2$. Proceeding as above we get

$$\int_\epsilon^\infty \frac{f(t)}{t^2}\,dt = \int_1^\infty \frac{f(t)}{t^2}\,dt + \int_\epsilon^1 \left[\frac{f(0)}{t^2} + \frac{f'(0)}{t} + \frac{f''(\theta t)}{2}\right]dt$$

$$= \int_1^\infty \frac{f(t)}{t^2}\,dt + \int_\epsilon^1 \frac{f''(\theta t)}{2}\,dt - f(0) + \left[\frac{f(0)}{\epsilon} - f'(0)\log\epsilon\right]$$

so that

$$\text{Fp}\int_{-\infty}^{+\infty} f(t)\frac{u(t)}{t^2}\,dt = \lim_{\epsilon\downarrow 0}\left[\int_1^\infty \frac{f(t)}{t^2}\,dt - f(0) + \int_\epsilon^1 \frac{f''(\theta t)}{2}\,dt\right]$$

$$= \int_1^\infty \frac{f(t)}{t^2}\,dt - f(0) + \lim_{\epsilon\downarrow 0}\int_\epsilon^1 \frac{f(t)-f(0)-tf'(0)}{t^2}\,dt$$

which reduces on integration by parts to

$$\int_1^\infty \frac{f'(t)}{t}\,\mathrm{d}t + \lim_{\epsilon\downarrow 0}\left[\int_\epsilon^1 \frac{f'(t)-f'(0)}{t}\,\mathrm{d}t + \frac{f(\epsilon)-f(0)}{\epsilon}\right].$$

Using (7.15) it follows that

$$\int_{-\infty}^{+\infty} f(t)\,\mathrm{Pf}\left[\frac{u(t)}{t^2}\right]\mathrm{d}t = \int_1^\infty \frac{f'(t)}{t}\,\mathrm{d}t + \int_0^1 \frac{f'(t)-f'(0)}{t}\,\mathrm{d}t + f'(0)$$

$$= -\int_{-\infty}^{+\infty}\{-f'(t)\}\,\mathrm{Pf}\left[\frac{u(t)}{t}\right]\mathrm{d}t - \int_{-\infty}^{+\infty} f(t)\delta'(t)\,\mathrm{d}t.\qquad(7.16)$$

That is,

$$\mathrm{Pf}\left[\frac{u(t)}{t^2}\right] = -\frac{\mathrm{d}}{\mathrm{d}t}\,\mathrm{Pf}\left[\frac{u(t)}{t}\right] - \delta'(t).\qquad(7.17)$$

In general it can be shown that

$$\frac{\mathrm{d}}{\mathrm{d}t}\,\mathrm{Pf}\left[\frac{u(t)}{t^m}\right] = -\mathrm{Pf}\left[\frac{mu(t)}{t^{m+1}}\right] + \frac{(-1)^m}{m!}\delta^{(m)}(t).\qquad(7.18)$$

7.3 TEST FUNCTIONS

7.3.1 The various kinds of pseudo-function discussed in the preceding sections of this chapter, together with the delta functions and their derivatives treated earlier, indicate the wide variety of generalised functions which can be introduced. The need for some sort of systematic and unified treatment is obvious. To achieve this it is necessary to specify some basic class of ordinary functions with respect to which the characteristic properties of all the generalised functions under consideration are well defined. We shall refer to the members of this basic class as **test functions**.

So far as the delta function and all its translates are concerned it would be enough to stipulate that test functions be continuous. But this would not ensure that the sampling property which characterises the first derivative of the delta function is always well-defined, and so we are led to require that all test functions be (at least) continuously differentiable. When we take into account the sampling properties required for derivatives of the delta function of arbitrary order then it becomes apparent that for a basic, universal, class of test functions we must have functions which are **infinitely differentiable** (that is, each test function has continuous derivatives of all orders.) But even this is not enough. In order to define the generalised function associated with the unit step function, or to specify any of the various pseudo-functions introduced above, it is necessary to work in terms of integrals which are known to be convergent. And this in turn generally makes it necessary to use test functions which each vanish outside some finite interval. Accordingly we are led to consider a fundamental set, $\mathfrak{D}$, of ordinary functions which enjoy the following properties:

(i) each function f in $\mathfrak{D}$ is infinitely differentiable,
(ii) for each function f in $\mathfrak{D}$ we can find a corresponding finite, closed, interval $[a,b]$ outside which f vanishes identically; f is then sometimes said to have its **support** contained in $[a,b]$.

As examples of functions belonging to $\mathfrak{D}$ (and which are of considerable interest in their own right) we may cite the following:

for $n = 1, 2, 3, \ldots$, define the functions ψ_n by

$$\psi_n(t) = \begin{cases} \dfrac{n}{A} \exp\left(\dfrac{1}{n^2 t^2 - 1}\right), & \text{for } |t| < 1/n \\ 0 & , \text{ for } |t| \geqslant 1/n . \end{cases} \tag{7.19}$$

7.3.2 If f and g are members of $\mathfrak{D}$ then so also is the function h defined by

$$h(t) = f(t) + g(t) . \tag{7.20}$$

Similarly if f belongs to $\mathfrak{D}$ and if α is any real number then αf is also a member of $\mathfrak{D}$, where αf denotes the function whose value at each point t is $\alpha f(t)$.

Thus $\mathfrak{D}$ is closed under a certain additive operation and under multiplication by real numbers. In this respect it has an algebraic structure which is analogous to that displayed by the three-dimensional space of ordinary experience when we treat it as a space of **vectors**. Given any two points in ordinary space, say $\mathbf{x} = (x_1, x_2, x_3)$ and $\mathbf{y} = (y_1, y_2, y_3)$, then we define their **vector sum** $\mathbf{x} + \mathbf{y}$ to be the point (or vector) specified by

$$\mathbf{z} \equiv \mathbf{x} + \mathbf{y} = (x_1 + y_1, x_2 + y_2, x_3 + y_3) .$$

Again if α is any real number then we define the product $\alpha \mathbf{x}$ to be the point

$$\alpha \mathbf{x} = (\alpha x_1, \alpha x_2, \alpha x_3) .$$

It is easily confirmed that these operations of **vector addition** and **scalar multiplication** will have the following properties:

L1. Vector addition is associative, commutative, and distributive with respect to scalar multiplication.

$$\mathbf{x} + (\mathbf{y} + \mathbf{z}) = (\mathbf{x} + \mathbf{y}) + \mathbf{z} ; \quad \mathbf{x} + \mathbf{y} = \mathbf{y} + \mathbf{x} ;$$

$$\alpha(\mathbf{x} + \mathbf{y}) = \alpha \mathbf{x} + \alpha \mathbf{y} .$$

L2. For any vector x and any scalars α and β we have

$$(\alpha + \beta)x = \alpha x + \beta x \; ; \quad \alpha(\beta x) = (\alpha\beta)x \;.$$

L3. There is a unique vector **0**, called the **null vector**, which is such that

$$x + 0 = 0 + x = x \quad \text{for every vector } x \;.$$

L4. To each vector x there corresponds a unique inverse $(-x)$ which is such that

$$x + (-x) = 0 \;.$$

L5. For every vector x it is the case that

$$1x = x \quad \text{and} \quad 0x = 0 \;.$$

In general any set of objects which is closed under an operation of combination (called vector addition) and under a form of multiplication by real, or by complex, numbers such that the conditions L1–L5 are satisfied, is called a **linear space** and its members are called **vectors**. When the scalar multipliers are confined to the real number system the space is called a **real linear space**; otherwise it is called a **complex linear space**. It is clear that the set of test functions $\mathfrak{D}$ with the definitions of pointwise sum (7.20) and of multiplication by real numbers given above does constitute a real linear space in this sense. Note that we could always obtain a corresponding complex linear space by considering the set of all complex-valued functions f where.

$$f(t) = f_1(t) + if_2(t)$$

and f_1 and f_2 are functions belonging to the real linear space $\mathfrak{D}$.

7.3.3 We can push the analogy between $\mathfrak{D}$ and ordinary three-dimensional space a stage further. If $x = (x_1, x_2, x_3)$ then we would normally understand the length of the vector x to be given by the number

$$\|x\| = +\sqrt{\{x_1^2 + x_2^2 + x_3^2\}} \;.$$

Conceptually this is the same as the "distance from the origin" of the point whose coordinates are x_1, x_2, and x_3. We can go on to speak of the distance between two arbitrary points, say (x_1, x_2, x_3) and (y_1, y_2, y_3) as the length of the vector $x - y$:

$$d(x,y) \equiv \|x - y\| = +\sqrt{\{(x_1 - y_1)^2 + (x_2 - y_2)^2 + (x_3 - y_3)^2\}} \;.$$

The concept of distance allows us to talk meaningfully about convergence and limiting processes in space in general, since there will be a specific sense in which we can speak of a sequence of points (or vectors) approaching more and more closely to some limiting point.

Similar ideas of distance and convergence can be developed for the space $\mathfrak{D}$ of infinitely differentiable functions. However, to do this it is necessary to isolate the essential formal properties which characterise "length" and "distance". We begin with a definition of **norm**.

A norm is a real-valued function $\|x\|$, defined on a linear space E, which satisfies the following conditions:

N1. $\|x\| \geqslant 0$ for every x in E, and $\|x\| = 0$ if and only if x is the null vector of E.
N2. $\|x + y\| \leqslant \|x\| + \|y\|$, (the "triangle" inequality).
N3. $\|\alpha x\| = |\alpha| \|x\|$, for every vector x and any scalar α.

A linear space E, together with a norm $\|x\|$ defined on E, is usually referred to as a **normed linear space**. The distance, $d(x,y)$, between any two members x and y of E is then defined as

$$d(x,y) = \|x - y\|$$

and the real-valued function $d(x,y)$ is spoken of as a **metric** on E. A metric can be defined from first principles as a real-valued function satisfying the conditions

D1. $d(x,y) \geqslant 0$ for every x and y, and $d(x,y) = 0$ if and only if $x = y$.
D2. $d(x,y) = d(y,x)$.
D3. $d(x,z) \leqslant d(x,y) + d(y,z)$.

Any set of objects on which such a function $d(x,y)$ is defined is said to be a **metric space**. Clearly every linear space with a norm is automatically a metric space with the metric defined in terms of the norm as explained above.

Once a norm has been defined on a linear space there is at once a corresponding sense of convergence in that space. In a metric space in general a sequence of points (x_m) is said to converge, in the sense of the metric on the space, to the point x as its limit if, given any number $\epsilon > 0$, we have

$$d(x, x_m) < \epsilon$$

for all but a finite number of values of m. That is to say, x is the limit of the x_m if and only if the distance between x and the x_m tends to zero as m tends to infinity:

$$\lim_{m \to \infty} x_m = x \quad \text{if and only if} \quad \lim_{m \to \infty} d(x, x_m) = 0 .$$

When, as here, the metric is defined in terms of a norm, the criterion for convergence becomes

$$\lim_{m \to \infty} \|x - x_m\| = 0 .$$

Convergence in this mode is often referred to as "convergence in norm".

7.3.4 There are many ways in which we might think of defining a norm, and hence a specific mode of convergence, on the linear space $\mathcal{D}$. To begin with, consider the so-called **uniform**, or **supremum**, norm

$$\|f\| = \sup_t |f(t)| . \tag{7.21}$$

The fact that (7.21) does constitute a norm is easy to show. In the first place it is clear that $\|f\| \geqslant 0$ for every f in $\mathfrak{D}$ and that $\|f\| = 0$ if and only if $f(t) = 0$ for all t. Also,

$$\|\alpha f\| = \sup_t |\alpha f(t)| = |\alpha| \sup_t |f(t)| = |\alpha| \|f\|$$

which establishes N3. Finally, the triangle inequality N2 follows since

$$|f(t) + g(t)| \leqslant |f(t)| + |g(t)| \quad \text{for each } t ,$$

so that

$$\sup \{|f(t) + g(t)|\} \leqslant \sup \{|f(t)| + |g(t)|\}$$

$$\leqslant \sup |f(t)| + \sup |g(t)| .$$

Now suppose that (f_m) is any sequence of functions in $\mathfrak{D}$ which converges to a limit f in the sense of the norm (7.21). Then

$$\lim_{m \to \infty} \|f - f_m\| = \lim_{m \to \infty} \{\sup |f(t) - f_m(t)|\} = 0$$

and the convergence is seen to be necessarily *uniform*. Note that this does not necessarily mean that the limit function f is itself a member of $\mathfrak{D}$. The uniformity of the convergence ensures that f is certainly bounded and continuous but not that it is differentiable. We cannot even conclude that f has support contained in some finite interval, but only that $|f(t)|$ tends to zero as $|t|$ becomes arbitrarily large:

$$\lim_{|t| \to \infty} |f(t)| = 0 .$$

On the other hand if it is the case that each of the functions f_m vanishes identically outside the same fixed interval $[a,b]$ then it is obvious that the limit function f must also vanish outside $[a,b]$. To distinguish this special case when it does arise we shall refer to the convergence of the f_m as **locally restricted uniform convergence**.

7.3.5 Instead of the uniform norm on $\mathfrak{D}$ we might choose instead to define a norm $\|f\|^{(1)}$ as follows:

$$\|f\|^{(1)} = \max \{\|f\|, \|f'\|\} = \max \{\sup_t |f(t)|, \sup_t |f'(t)|\} . \qquad (7.22)$$

Convergence in the sense of this norm is stronger than ordinary uniform convergence. For if a sequence (f_m) converges to a limit f in the sense of (7.22) then we must have both

$$\lim_{m \to \infty} \text{-}[\sup |f(t) - f_m(t)|] = 0$$

and

$$\lim_{m \to \infty} \{\sup |f'(t) - f_m'(t)|\} = 0 .$$

That is, (f_m) converges uniformly to f and, in addition, the sequence (f'_m) of the first derivatives of the f_m converges uniformly to the limit f'. Once again convergence in this sense does not necessarily ensure that the limit f is a member of $\mathfrak{D}$. In general we can only conclude that f must always be at least continuously differentiable and that both $|f(t)|$ and $|f'(t)|$ tend to zero as $|t|$ tends to infinity.

More generally, for any given positive integer p, we can define a norm $\|f\|^{(p)}$ on $\mathfrak{D}$ by writing

$$\|f\|^{(p)} = \max \{\|f\|, \|f'\|, \ldots, \|f^{(p)}\|\}$$
$$= \max_{0 \leqslant k \leqslant p} \{\sup |f^{(k)}(t)|\}. \tag{7.23}$$

The mode of convergence appropriate to this norm is most conveniently referred to as **p-uniform convergence**. A sequence (f_m) is said to converge p-uniformly to the limit function f if and only if each of the sequences $(f_m^{(k)})$ of the kth derivatives of the f_m (for $k = 0, 1, \ldots, p$) converges uniformly to the corresponding derivative $f^{(k)}$ of the limit function f. We shall describe the convergence as **locally restricted p-uniform convergence** if, in addition, each of the functions f_m vanishes identically outside the same fixed interval $[a,b]$.

Armed with these possible metrical structures on the linear space $\mathfrak{D}$ we can now consider the characterisation of the generalised functions encountered so far. Each of these has been recognised as essentially a mapping, defined on an appropriate class of ordinary functions and taking numerical values. In the next section we shall be more specific about the nature and the properties of such mappings, particularly in relation to the algebraic structure of the basic space of test functions $\mathfrak{D}$ and to the various modes of convergence which can be defined within it.

7.4 LINEAR FUNCTIONALS AND DISTRIBUTIONS

7.4.1 Given a linear space E let μ denote a mapping which assigns to each member f of E a certain well-defined number $\mu(f)$. Any such numerically valued function on E is called a **functional**. We say that μ is a **linear functional** on E if it is the case that

$$\mu(\alpha f + \beta g) = \alpha\mu(f) + \beta\mu(g) \tag{7.24}$$

for every f and g in E and every pair of scalars (numbers) α and β. If, in addition, E is a normed linear space and there exists some fixed, finite, positive number K such that

$$|\mu(f)| \leqslant K\|f\| \tag{7.25}$$

for every f in E, then μ is said to be a **bounded** linear functional on E. Every bounded linear functional on E is necessarily continuous in the sense that whenever a sequence (f_m) in E converges in norm to a limit f in E then the corresponding numerical sequence $(\mu(f_m))$ converges in the usual sense to the number $\mu(f)$. This follows at once since we have

$$|\mu(f_m) - \mu(f)| = |\mu(f - f_m)| \leqslant K \|f - f_m\|$$

and, by hypothesis, $\|f - f_m\|$ tends to 0 as m tends to infinity.

The set of all bounded linear functionals on a given linear space E turns out to be a normed linear space in its own right, with the following definitions of vector addition, scalar multiplication, and norm:

Addition. The sum $(\mu_1 + \mu_2)$ of two bounded linear functionals μ_1 and μ_2 is the bounded linear functional μ defined by

$$\mu(f) = \mu_1(f) + \mu_2(f) , \quad \text{for every } f \text{ in } E .$$

Scalar multiplication. The product $\alpha\mu$, where μ is a bounded linear functional on E and α is a scalar, is defined by

$$(\alpha\mu)(f) = \alpha\mu(f) , \quad \text{for every } f \text{ in } E .$$

Norm. We define the norm of a bounded linear functional μ to be

$$\|\mu\| = \sup_{\|f\| \leqslant 1} |\mu(f)| . \tag{7.26}$$

It is easy to see that the above definitions of addition and scalar multiplication do satisfy the required conditions for the operations of a linear space, and it is only slightly more difficult to establish that $\|\mu\|$ is a true norm.

In what follows we shall be primarily concerned with real linear functionals defined on real linear spaces; however, it is a trivial exercise to extend the results to the case of complex-valued functionals defined on real, or complex, linear spaces.

7.4.2 Consider first the linear space $\mathfrak{D}$ with the metric structure imposed by the uniform norm

$$\|f\| = \sup_t |f(t)| .$$

Let μ be a linear functional defined on $\mathfrak{D}$ which is bounded with respect to this norm. That is to say we assume that $\mu(f)$ is well-defined for every function f belonging to $\mathfrak{D}$ and that there exists some finite number K such that

$$|\mu(f)| \leqslant K \|f\| = K \sup |f(t)| \tag{7.27}$$

for all f in $\mathfrak{D}$.

Examples

(i) Let h be any fixed function which is absolutely integrable over the whole range $-\infty < t < +\infty$. That is, h is such that

$$\int_{-\infty}^{+\infty} |h(t)| \, dt < +\infty .$$

Then the mapping

$$f \to \int_{-\infty}^{+\infty} f(t) h(t) \, dt$$

is well-defined for all f in $\mathfrak{D}$ and is plainly linear. Moreover we have

$$\left| \int_{-\infty}^{+\infty} f(t) h(t) \, dt \right| \leq \int_{-\infty}^{+\infty} |f(t)| \, |h(t)| \, dt \leq \sup |f(t)| \int_{-\infty}^{+\infty} |h(t)| \, dt$$

so that the mapping is bounded with respect to the uniform norm. If we write

$$\mu(f) \equiv \int_{-\infty}^{+\infty} f(t) h(t) \, dt$$

then it follows that μ is a linear functional on $\mathfrak{D}$, bounded with respect to the uniform norm, and that

$$\|\mu\| = \int_{-\infty}^{+\infty} |h(t)| \, dt . \qquad (7.28)$$

(ii) For any fixed real number a the mapping $f \to f(a)$ clearly defines a linear functional on $\mathfrak{D}$ which is bounded with respect to the uniform norm. We shall denote this functional by the symbol δ_a and refer to it as the **Dirac delta functional located at $t = a$**:

$$\left.\begin{array}{c} \delta_a(f) = f(a) \\ \|\delta_a\| = 1 \end{array}\right\} \qquad (7.29)$$

The functional δ_a carries out the sampling operation hitherto assigned to the translate $\delta(t - a)$ of the delta function; where convenient, however, we shall continue to use the (purely symbolic) notation:

$$\delta_a(f) \equiv \int_{-\infty}^{+\infty} f(t) \delta(t - a) \, dt = f(a) .$$

(iii) Finite linear combinations of delta functions and absolutely integrable functions can be represented in the same way in terms of linear functionals on $\mathfrak{D}$, bounded with respect to the uniform norm. However, we need an extension of these ideas if we are to obtain like representations of the generalised function defined by the unit step function u (which is not absolutely integrable over $(-\infty, +\infty)$) or of infinite combinations of delta functions such as

$$\delta(\sin t) \equiv \sum_{m=-\infty}^{+\infty} \delta(t - m\pi) \;.$$

Consider again the linear functional defined by the mapping

$$f \to \int_{-\infty}^{+\infty} f(t)h(t)\,\mathrm{d}t$$

where h is now not necessarily absolutely integrable over $(-\infty, +\infty)$ but merely *locally integrable*; that is, such that

$$\int_{a}^{b} |h(t)|\,\mathrm{d}t < +\infty$$

for any finite interval $[a,b]$.

Given any f in $\mathfrak{D}$ there must exist some finite interval $[a,b]$ outside which f vanishes identically. Hence

$$\left| \int_{-\infty}^{+\infty} f(t)h(t)\,\mathrm{d}t \right| = \left| \int_{a}^{b} f(t)h(t)\,\mathrm{d}t \right| \leqslant \|f\| \int_{a}^{b} |h(t)|\,\mathrm{d}t \;.$$

In general we say that a linear functional μ is **relatively** (or, **locally**) bounded with respect to the uniform norm on $\mathfrak{D}$ if, for each finite real interval $I = [a,b]$, we can find a finite positive number $K(I)$ such that

$$|\mu(f)| \leqslant K(I) \sup |f(t)| \tag{7.30}$$

for every function f in $\mathfrak{D}$ with support contained in I. In the above example we have

$$K(I) = \int_{a}^{b} |h(t)|\,\mathrm{d}t \;.$$

Again, if

$$\mu = \delta(\sin t) = \sum_{m=-\infty}^{+\infty} \delta(t - m\pi)$$

then for any finite interval $[a,b]$ there will exist integers n_1, n_2, such that

$$(n_1 - 1)\pi \leqslant a < b \leqslant (n_2 + 1)\pi \;.$$

Hence for any f in $\mathcal{D}$ with support contained in $I = [a,b]$ we have

$$|\mu(f)| = \left| \sum_{m=n_1}^{n_2} f(m\pi) \right| \leqslant \sum_{m=n_1}^{n_2} |f(m\pi)| \leqslant (n_2 - n_1 + 1) \|f\|$$

so that μ is relatively bounded with respect to the uniform norm.

Note that boundedness always implies relative boundedness but not conversely. A relatively bounded linear functional on $\mathcal{D}$ will actually be a bounded linear functional on $\mathcal{D}$ only if the numbers $K(I)$ in (7.30) are themselves uniformly bounded; that is, only if there exists some finite number K such that $K(I) \leqslant K$ for all intervals I.

7.4.3 Let μ be any linear functional on $\mathcal{D}$ which is relatively bounded for the uniform norm. We denote by μ' the linear functional on $\mathcal{D}$ which is defined by

$$\mu'(f) = \mu(-f') \tag{7.31}$$

for all f in $\mathcal{D}$; μ' will be referred to as the **derivative** of μ. If in particular μ is defined by

$$\mu(f) = \int_{-\infty}^{+\infty} f(t)h(t)\mathrm{d}t \, ,$$

where h is a locally integrable function which has a locally integrable derivative h' in the classical sense, then an integration by parts shows that for any f in $\mathcal{D}$,

$$\int_{-\infty}^{+\infty} f(t)h'(t)\mathrm{d}t = \int_{-\infty}^{+\infty} \{-f'(t)\}h(t)\mathrm{d}t \, .$$

That is, this notion of derivative is compatible with differentiation as used in ordinary calculus.

In general, if f is any function in $\mathcal{D}$ which vanishes outside an interval $I = [a,b]$, then

$$|\mu'(f)| = |\mu(-f')| \leqslant K(I) \sup |f'(t)| \leqslant K(I) \|f\|^{(1)} \, .$$

Thus μ' will certainly always be relatively bounded with respect to the norm $\|f\|^{(1)}$, though not necessarily with respect to the uniform norm $\|f\|$ itself.

Examples

(i) The derivative of δ_a is defined by the mapping

$$f \to \delta_a(-f') = -f'(a) \, .$$

Then, trivially,

$$|\delta_a'(f)| = |f'(a)| \leqslant \|f'\| \leqslant \|f\|^{(1)} \, .$$

On the other hand given any positive number N, however large, we can always find some f in $\mathfrak{D}$ such that $|f(t)| \leqslant 1$ for all t, but $|f'(a)| > N$. This enough to show that δ_a' is not relatively bounded with respect to the uniform norm.

(ii) The locally integrable function $u(t) \log |t|$ has the (generalised) derivative $\mathrm{Pf}\{u(t)/t\}$. For any function f in $\mathfrak{D}$ it is the case that

$$\int_{-\infty}^{+\infty} f(t)\, \mathrm{Pf}\left[\frac{u(t)}{t}\right] \mathrm{d}t \equiv \mathrm{Fp}\int_{-\infty}^{+\infty} f(t)\frac{u(t)}{t}\,\mathrm{d}t = \int_{-\infty}^{+\infty} \{-f'(t)\}\, u(t) \log |t|\,\mathrm{d}t$$

and it follows that the functional defined by the pseudo-function $\mathrm{Pf}\{u(t)/t\}$ is the derivative, in the sense of (7.31), of the functional defined by the locally integrable function $u(t) \log |t|$. What is more, we have

$$\left|\int_{-\infty}^{+\infty} f(t)\, \mathrm{Pf}\left[\frac{u(t)}{t}\right]\mathrm{d}t\right| = \left|\int_{0}^{\infty} \frac{f(t)-f(0)}{t}\,\mathrm{d}t\right| = \left|\int_{0}^{b} f'(\theta t)\mathrm{d}t\right| \leqslant b \sup |f'(t)|$$

for some b and some θ such that $0 < \theta < 1$.

Hence the functional concerned is relatively bounded with respect to the norm $\|f\|^{(1)}$.

7.4.4 By applying the differentiation process defined by (7.31) to linear functionals relatively bounded with respect to the norm $\|f\|^{(1)}$ we would similarly generate linear functionals relatively bounded with respect to the norm $\|f\|^{(2)}$, and so on. This leads us to suggest a tentative definition of a generalised function as any linear functional on $\mathfrak{D}$ which is relatively bounded with respect to some norm $\|f\|^{(p)}$, where $p = 0, 1, 2, \ldots$. Following Schwartz we call such a functional a **distribution**; strictly, we should refer to it as a distribution of **finite order**, because the definition used by Schwartz is more general and includes linear functionals other than the ones discussed so far.

7.5 THE CALCULUS OF DISTRIBUTIONS

7.5.1 Any distribution defined by a locally integrable function h is said to be a **regular** distribution; in many contexts we can identify the function h with the functional μ which it defines, though, as shown below, this must be done with due caution. Any other kind of distribution is said to be **singular**. From what has been said before about functionals in general it is a straightforward matter to summarise the algebra appropriate to distributions.

(i) *Equality.* Two distributions μ_1 and μ_2 are said to be equal if and only if $\mu_1(f)$ and $\mu_2(f)$ are equal for every f in $\mathfrak{D}$. Note that if μ_1 and μ_2 are regular distributions defined by locally integrable functions h_1 and h_2 respectively then

$\mu_1 = \mu_2$ does not necessarily imply that $h_1(t) = h_2(t)$ for all t. The most we can say is that

$$\int_a^b |h_1(t) - h_2(t)| \, dt = 0$$

for every finite interval $[a,b]$. This means that h_1 and h_2 can differ in value at most on the points of a set which, in a sense which can be made quite precise, is "negligibly small". (Roughly speaking, the set of points t for which $h_1(t) \neq h_2(t)$ might be described as of zero total length). It is usual to say that $h_1(t) = h_2(t)$ "almost everywhere".

(ii) The **null distribution 0** is the linear functional which maps every function f in $\mathfrak{D}$ into zero:

$$\mathbf{0}(f) = 0 \quad \text{for every } f \text{ in } \mathfrak{D}.$$

It is a regular distribution which can be defined by, or represented by, any function h which vanishes almost everywhere; that is, any function h such that

$$\int_a^b |h(t)| \, dt = 0$$

for every finite interval $[a,b]$.

(iii) **Addition** and **scalar multiplication** for distributions are defined as for linear functionals in general. That is to say we define

$$(\mu_1 + \mu_2)(f) \equiv \mu_1(f) + \mu_2(f)$$

$$(\alpha\mu)(f) \equiv \alpha \{\mu(f)\}$$

for all f in $\mathfrak{D}$, where μ_1, μ_2, and μ are distributions and α is any scalar. In particular note that for any distribution μ we get $\mu + \mathbf{0} = \mathbf{0} + \mu = \mu$ and $\mathbf{0}(\mu) = \mathbf{0}$.

(iv) A function ϕ is called a **multiplier** for a distribution μ if $\mu(\phi f)$ is well-defined for every function f for which $\mu(f)$ exists. If ϕ is a multiplier for μ then the product $\phi\mu$ is defined to be the distribution given by $(\phi\mu)(f) \equiv \mu(\phi f)$ for every f in $\mathfrak{D}$. For a general distribution μ the product $\phi\mu$ is defined for any ϕ in $\mathfrak{D}$. For particular distributions the multiplier ϕ may be considerably less restricted. Thus if ϕ is merely assumed to be continuous then, as we saw in Chapter 3,

$$(\phi\delta)(f) = \phi(0)f(0) \equiv \phi(0)\delta(f).$$

Similarly, provided only that ϕ is continuously differentiable, we can define the product $\phi\delta'$ by

$$(\phi\delta')(f) \equiv \delta'(\phi f) = -\phi'(0)f(0) - \phi(0)f'(0)$$

$$\equiv -\phi'(0)\delta(f) + \phi(0)\delta'(f).$$

In the case of regular distributions there is obviously a close correspondence between the algebraic operations carried out on the distributions themselves and those carried out on the locally integrable functions which define those distributions. Now we cannot strictly identify a regular distribution μ with a locally integrable function h. μ can be identified only with a certain **equivalence class** $\{h\}$ of functions, any two members h_1, h_2, of which will be equal almost everywhere. The regular distribution associated with the unit step function, for example, could equally well be defined by any one of the locally integrable functions $u_c(t)$, but should not be identified with any one of them.

The importance of distinguishing between the functions and processes of ordinary calculus and the generalised functions, or distributions, and the corresponding operations on them is most clearly shown in the case of products involving distributions. Unless some restrictions are imposed on the factors the associative law may fail. For example, the product $t\delta$ is plainly the null distribution, $\mathbf{0}$, and so we should have

$$\frac{1}{t}(t\delta) = \mathbf{0} .$$

On the other hand it seems equally clear that

$$\left(\frac{1}{t}t\right)\delta = 1\delta = \delta .$$

7.5.2 *Convergence.* A sequence (μ_n) of distributions is said to converge to the distribution μ as its limit if and only if

$$\mu(f) = \lim_{n \to \infty} \mu_n(f)$$

for every f in $\mathfrak{D}$.

The infinite series of distributions

$$\sum_{n=1}^{\infty} \mu_n$$

converges to the sum μ if and only if the sequence of partial sums

$$\sum_{k=1}^{n} \mu_k$$

converges to μ as its limit; that is to say

$$\mu = \sum_{n=1}^{\infty} \mu_n \quad \text{if and only if} \quad \mu = \lim_{n \to \infty} \sum_{k=1}^{n} \mu_k .$$

The most familiar context in which the convergence of distributions arises is that of the definition of the delta function as the formal limit of sequences of ordinary functions (see Chapter 2). In general let k be some fixed, absolutely integrable, function; without loss of generality we may suppose that

$$(a) \ k(t) \geqslant 0 \text{ for all } t, \quad \text{and} \quad (b) \int_{-\infty}^{+\infty} k(t) dt = 1 .$$

Now denote by μ_n the regular distribution given by

$$\mu_n(f) = \int_{-\infty}^{+\infty} k_n(t) dt , \quad \text{where} \quad k_n(t) \equiv nk(nt) .$$

Then for f in $\mathcal{D}$ we have

$$\mu_n(f) = \int_{-\infty}^{+\infty} f(\tau) k_n(\tau) d\tau = \int_{-\infty}^{+\infty} f(t/n) k(t) dt$$

$$= \int_{-\infty}^{+\infty} f(0) k(t) dt + \int_{-\infty}^{+\infty} \{f(t/n) - f(0)\} k(t) dt$$

Also,

$$\left| \int_0^{\infty} \{f(t/n) - f(0)\} k(t) dt \right|$$

$$\leqslant \int_0^{R} |f(t/n) - f(0)| \, |k(t)| dt + \int_R^{\infty} |f(t/n) - f(0)| \, |k(t)| dt$$

$$= I_1 + I_2 \quad \text{say.}$$

For all sufficiently large R we have (since k is absolutely integrable)

$$I_2 \leqslant 2\|f\| \int_R^{\infty} |k(t)| dt < \epsilon/2 .$$

Having chosen R, note that if $0 \leqslant t \leqslant R$ then $0 \leqslant t/n \leqslant R/n$. Hence by choosing n sufficiently large we can ensure that

$$\sup_{0 \leqslant t \leqslant R} |f(t/n) - f(0)| < \epsilon/2R .$$

A similar argument applies to

$$\int_{\infty}^{0} \{f(t/n) - f(0)\} k(t) dt ,$$

and it follows that (μ_n) converges to δ in the sense that

$$\lim_{n \to \infty} \mu_n(f) \equiv \lim_{n \to \infty} \int_{-\infty}^{+\infty} f(t)k_n(t)\,dt = f(0), \quad \text{for all } f \text{ in } \mathfrak{D}^\dagger.$$

A sequence of functions of the form (k_n) is often called a **delta sequence**. The following examples of delta sequences are of particular importance:

(i) $k(t) = 1/\pi(t^2 + 1)$ $\delta(t) = \lim_{n \to \infty} n/\pi(n^2 t^2 + 1)$

(ii) $k(t) = \dfrac{1}{2}\exp(-|t|)$ $\delta(t) = \lim_{n \to \infty} \dfrac{n}{2}\exp(-|nt|)$

(iii) $k(t) = \dfrac{1}{\sqrt{\pi}}\exp(-t^2)$ $\delta(t) = \lim_{n \to \infty} \dfrac{n}{\sqrt{\pi}}\exp(-n^2 t^2)$

(iv) $k(t) = \dfrac{1}{\pi}\left(\dfrac{\sin t}{t}\right)^2$ $\delta(t) = \lim_{n \to \infty} \dfrac{n}{\pi}\left(\dfrac{\sin nt}{t}\right)^2$

7.6 DISTRIBUTIONS IN THE SENSE OF SCHWARTZ

The theory of generalised functions outlined in this chapter makes use only of the more accessible parts of modern analysis associated with the elementary theory of normed linear spaces. The theory of distributions, as developed by Laurent Schwartz, is more general than the account given here and demands rather more sophisticated mathematical apparatus. Briefly, suppose we take the norm $\|f\|^{(p)}$ defined by equation (7.23), and the corresponding concept of p-uniform convergence, and then allow p to tend to infinity. On the one hand we cannot define a norm $\|f\|^\infty$ on $\mathfrak{D}$ which is in any sense the "limit" of the norm $\|f\|^{(p)}$; on the other, however, the concept of p-uniform convergence can be extended in a meaningful way.

A sequence (f_n) in $\mathfrak{D}$ may be said to converge ω-**uniformly** to the limit f if and only if each of the sequences $(f_n^{(k)})$ of the kth derivatives of the f_n (for $k = 0, 1, 2, \ldots$) converges uniformly to the kth derivative of f; if all the f_n vanish outside the same fixed interval $[a,b]$ then the limit function f also vanishes outside $[a,b]$ and the convergence is said, as usual, to be locally restricted.

With this sense of convergence the linear space $\mathfrak{D}$ has a structure which is akin to, but more general than, that defined by a norm or a metric. Any linear functional μ on $\mathfrak{D}$ is said to be a **distribution in the sense of Schwartz** if it is

---

† The proof given here actually shows that the limit is $f(0)$ for any bounded, continuous function f.

continuous under locally restricted ω-uniform convergence on $\mathfrak{D}$; that is, if we have

$$\mu(f) = \lim_{n \to \infty} \mu(f_n)$$

for every sequence (f_n) of functions in $\mathfrak{D}$ such that

(i) for $k = 0, 1, 2, \ldots$, each sequence $(f_n^{(k)})$ converges uniformly to the limit $f^{(k)}$, and

(ii) there exists some finite interval $[a,b]$ outside which each of the functions f_n vanishes identically.

Plainly, if μ is any linear functional which is relatively bounded on $\mathfrak{D}$ with respect to some norm $\|f\|^{(p)}$, where $p = 0, 1, 2, \ldots$, then μ is certainly continuous under locally restricted ω-uniform convergence on $\mathfrak{D}$. That is to say, all the generalised functions which we have studied so far are distributions in the sense of Schwartz. Distributions like the delta function itself, which are relatively bounded on $\mathfrak{D}$ with respect to the uniform norm, are often referred to as **distributions** of **order 0**. Distributions like the first derivative, δ', of the delta function, which are relatively bounded on $\mathfrak{D}$ with respect to the norm $\|f\|^{(1)}$, but not with respect to the uniform norm $\|f\|$, are similarly referred to as **distributions** of **order 1**, and so on. In practice it is these distributions of finite order which turn out to be the most generally useful. The distributions of order 0 occupy a position of central importance in the theory, not least because of their intimate connection with integration. This point is explored further in Chapter 8.

CHAPTER 8

Generalised Functions and Integration Theory

8.1 THE RIEMANN–STIELTJES INTEGRAL

8.1.1 Within the framework of the theory of distributions outlined in Chapter 7 the delta function is defined as a linear functional only in so far as infinitely differentiable functions are concerned. In the main body of the text, however, we have seen that it is useful to have the basic sampling property of the delta function defined with respect to functions for which we assume nothing stronger than simple continuity on some neighbourhood of the origin. Some extension is possible within the framework of the theory of distributions itself. However, a more thoroughgoing treatment of the problem, and one which turns out to be of very considerable interest and importance in its own right, is provided by the Stieltjes integral representation touched on in Chapter 2. The elementary theory of the Stieltjes integral (which we describe below) is adequate for continuous integrands but runs into difficulties in the case of discontinuous integrands. Hence, if we are to use Stieltjes integration to give us a means of extending the delta function sampling property we will need a more sophisticated theory of integration than the elementary, or **Riemann–Stieltjes**, theory. In order to see just what the elementary theory does allow, and where the difficulties arise, we begin by sketching the basic theory in rather more detail than in Chapter 2.

8.1.2 As usual we consider integration over a finite closed interval $[a,b]$. Let v be some fixed, monotone increasing function, hereinafter called the **integrator**, and f an arbitrary bounded function, called the **integrand**. As in Chapter 2 we shall define a partition, P, of $[a,b]$ to be a sub-division of that interval by finitely many points t_k, where $0 \leqslant k \leqslant n$, and

$$a = t_0 < t_1 < t_2 < \ldots < t_n = b .$$

Then for each given partition P of $[a,b]$ we can always form the following sums:

Lower sum. $\quad s(f,v,P) \equiv \sum_{k=1}^{n} m_k \{v(t_k) - v(t_{k-1})\} \equiv \sum_{k=1}^{n} m_k \Delta_k v$

Upper sum. $S(f,v,P) \equiv \sum\limits_{k=1}^{n} M_k \{v(t_k) - v(t_{k-1})\} \equiv \sum\limits_{k=1}^{n} M_k \Delta_k v$

where $M_k = \sup f(t)$ and $m_k = \inf f(t)$ for $t_{k-1} \leqslant t \leqslant t_k$, and we write

$$\Delta_k v \equiv v(t_k) - v(t_{k-1}) \, .$$

Clearly, it is always the case that

$$m[v(b) - v(a)] \leqslant s(f,v,P) \leqslant S(f,v,P) \leqslant M[v(b) - v(a)]$$

where $M = \sup f(t)$ and $m = \inf f(t)$ for $a \leqslant t \leqslant b$.

Thus, as P ranges over all possible partitions of $[a,b]$, the set $\{s\}$ of all possible lower sums is bounded above and so has a least upper bound; similarly the set $\{S\}$ of all upper sums is bounded below and so has a greatest lower bound. As a result we can always define the so-called lower and upper Riemann–Stieltjes integrals of f with respect to v,

$$\underline{\int_a^b} f(t) \mathrm{d}v(t) = \sup_P \{s(f,v,P)\} \; ; \quad \overline{\int_a^b} f(t) \mathrm{d}v(t) = \inf_P \{S(f,v,P)\}$$

It can be shown that the Riemann–Stieltjes lower integral is always less than or equal to the Riemann–Stieltjes upper integral:

$$\underline{\int_a^b} f(t) \mathrm{d}v(t) \leqslant \overline{\int_a^b} f(t) \mathrm{d}v(t) \, . \tag{8.1}$$

If these two semi-integrals are actually equal then the function f is said to be integrable in the Riemann–Stieltjes sense with respect to v over the range $[a,b]$. The Riemann–Stieltjes (RS) integral of f with respect to v over $[a,b]$ is then taken to be the common value of the upper and lower integrals.

8.1.3 A necessary and sufficient condition for the Riemann-Stieltjes integrability of f with respect to v is that, given any $\epsilon > 0$, we can always find a corresponding partition P_ϵ for which it is the case that $S - s < \epsilon$. Using this criterion we can establish a key result for the elementary theory of the integral, namely, that every function f continuous on $[a,b]$ must necessarily be Riemann–Stieltjes integrable over $[a,b]$ with respect to any monotone increasing integrator v.

To see this, note first that any function f which is continuous on a finite closed interval $[a,b]$ must actually be **uniformly** continuous there. That is to say, given any $\epsilon > 0$, we can always find a corresponding positive number $\eta = \eta(\epsilon)$ such that

$$|f(t) - f(\tau)| < \epsilon$$

for all points t and τ in $[a,b]$ such that $|t - \tau| < \eta$. Now choose a partition P_ϵ of $[a,b]$ such that

$$\Delta \equiv \max \, (t_k - t_{k-1}) < \eta \; .$$

For such a partition the corresponding sums S_ϵ and s_ϵ must be such that

$$S_\epsilon - s_\epsilon = \sum_{k=1}^{n} \{M_k - m_k\}\Delta_k v < \sum_{k=1}^{n} \epsilon \Delta_k v \; .$$

But

$$\sum_{k=1}^{n} \Delta_k v = \{v(t_1) - v(t_0)\} + \{v(t_2) - v(t_1)\} + \ldots$$

$$+ \{v(t_n) - v(t_{n-1})\} = v(b) - v(a) \; ,$$

so that

$$S_\epsilon - s_\epsilon < \epsilon[v(b) - v(a)] \; .$$

That is, $S_\epsilon - s_\epsilon$ can be made as small as we wish, and the result follows.

8.1.4 The limitations of the elementary theory can be readily seen from the following example. Using the notation introduced in Chapter 2 for unit step functions, suppose we try to evaluate the Stieltjes integral

$$\int_{-1}^{+1} u_{1/2}(t)\,\mathrm{d}u_c(t) \; . \tag{8.2}$$

For any partition of $[-1,+1]$ there can be only one of two possibilities

- (i) the origin is an interior point of some sub-interval, in which case $S = 1$ and $s = 0$,
- (ii) the origin is a boundary point of two adjacent sub-intervals; this time we have $S = c/2 + (1-c) = 1-c/2$, and $s = (1-c)/2$. Thus, for every partition of $[-1,+1]$, it will certainly be the case that

$$S - s \geqslant 1/2$$

and so the integral in question is undefined in the Riemann–Stieltjes sense. In particular this shows that it is not possible to use the elementary Riemann–Stieltjes integral to extend the sampling property of the delta function so that it applies to functions with simple discontinuities at the origin.

This difficulty with the integral of (8.2) stems, of course, from the fact that the integrator and the integrand have a common discontinuity at the origin. However, there are awkward features of the theory even when the integrator is a smooth function, so that the Stieltjes integral reduces to an ordinary Riemann integral. There is a classical example of this in the case of the sequence of functions f_m defined as follows:

$$f_m(t) = \lim_{n \to \infty} [\cos(m!\pi t)]^{2n} \quad \text{for } 0 \leqslant t \leqslant 1 .\tag{8.3}$$

Now $|\cos m!\pi t| = 1$ if $m!t$ is an integer and is strictly less than 1 otherwise. Hence, allowing n to tend to infinity, we find that $f_m(t) = 1$ for those values of t in $[0,1]$ for which $m!t$ is an integer and is zero for all other values of t. This means that $f_m(t)$ is non-zero for only finitely many values of t in $[0,1]$. For any partition P of $[0,1]$ we will always have a lower sum $s(f_m,t,P) = 0$. Further, the upper sum $S(f_m,t,P)$ can be made as small as we wish by choosing any partition with sufficiently small subdivisions. It follows that for each m the function f_m is integrable in the Riemann sense over $[0,1]$ and that we have

$$\int_0^1 f_m(t)\mathrm{d}t = 0 .$$

If $f(t) = \lim_{m \to \infty} f_m(t)$, then we now consider the integrability of f in the Riemann sense.

First, if t is irrational, then there exists no value of m for which $m!t$ is an integer; hence $f_m(t) = 0$ for every m and so $f(t) = 0$. On the other hand, if $t = p/q$ where p and q are integers then $m!t$ is an integer for all m such that $m \geqslant q$, and so $f_m(t) = 1$ for all such m and so $f(t) = 1$.

Since every sub-interval of $[0,1]$ must contain both rationals and irrationals it follows that for every partition of $[0,1]$ we must have $s(f,t,P) = 0$ and $S(f,t,P) = 1$. That is, the integral of f over the interval $[0,1]$ in the elementary, or Riemann, sense does not exist. It was primarily considerations of this kind which made a more sophisticated theory of integration desirable in the first place. The point is not so much the existence of functions like f which are non-integrable in terms of the elementary theory, but that such functions can arise as the limits of sequences of functions which are themselves all integrable. In what follows we shall sketch a treatment of the modern **Lebesgue** approach to integration theory in a form which applies not only to the integrals of ordinary analysis but to Stieltjes integrals in general.

8.2 EXTENSION OF THE ELEMENTARY STIELTJES INTEGRAL

8.2.1 From the elementary definition of the integral it is easy to establish that it enjoys the following properties:

(i) *Linearity.* If f_1 and f_2 are any two bounded functions each integrable over $[a,b]$ with respect to the monotonic increasing integrator v, and if α and β are any real numbers, then the function $f = \alpha f_1 + \beta f_2$ is also integrable over $[a,b]$ with respect to v and

$$\int_a^b f(t)\,dv(t) = \alpha \int_a^b f_1(t)\,dv(t) + \beta \int_a^b f_2(t)\,dv(t) . \qquad (8.4)$$

(ii) *Positiveness.* For any function f, integrable with respect to v over $[a,b]$, which is such that $f(t) \geqslant 0$ for all t in $[a,b]$ we have

$$\int_a^b f(t)\,dv(t) \geqslant 0 \qquad (8.5)$$

(iii) *Absoluteness.* If f is integrable over $[a,b]$ with respect to the monotone increasing function v then so also is the function $|f|$. Moreover we have

$$\left| \int_a^b f(t)\,dv(t) \right| = \left| \int_a^b f^+(t)\,dv(t) - \int_a^b f^-(t)\,dv(t) \right|$$

$$\leqslant \int_a^b f^+(t)\,dv(t) + \int_a^b f^-(t)\,dv(t) = \int_a^b |f(t)|\,dv(t)$$

where f^+ and f^- denote respectively the positive and negative components of f:

$$f^+(t) = \max\{f(t), 0\} ; \quad f^-(t) = \max\{-f(t), 0\} .$$

(iv) *Boundedness.* If f is integrable over $[a,b]$ with respect to the integrator v, and if $M = \sup |f(t)|$ for $a \leqslant t \leqslant b$, then

$$\left| \int_a^b f(t)\,dv(t) \right| \leqslant M[v(b) - v(a)] . \qquad (8.6)$$

Now let the integrator v be some fixed monotone increasing function and let f be an arbitrary function which is continuous everywhere and which vanishes identically outside some finite interval, say $I = [a,b]$. Then the Stieltjes integral of f with respect to v over the interval $[a,b]$ certainly exists and we may actually write

$$\int_a^b f(t)\,dv(t) = \int_{-\infty}^{+\infty} f(t)\,dv(t) < +\infty$$

without any ambiguity and without any need to discuss convergence of the (technically improper) integral on the right-hand side.

We can conveniently restate this result in the language of Chapter 7. First we can easily show that the set $C_0(I)$ of all continuous functions f which vanish outside the same fixed interval I is a linear space. Next it follows from (8.4) and (8.6) that the mapping

$$f \rightarrow \int_{-\infty}^{+\infty} f(t)\,\mathrm{d}\nu(t) \qquad (8.7)$$

defines a linear functional on the space $C_0(I)$ which is bounded with respect to the uniform norm $\|f\| = \sup_t |f(t)|$. Further, if we denote by C_0 the linear space of all continuous functions f each of which vanishes identically outside some finite interval (that is, C_0 denotes the union of the spaces $C_0(I)$ as I ranges over all possible finite closed intervals) then (8.7) is seen to define a linear functional on C_0 which is relatively bounded with respect to the uniform norm. Moreover, since the Schwartz space $\mathfrak{D}$ is contained in the space C_0, we may conclude that the linear functional defined by any mapping of the form (8.7) is necessarily a distribution — more precisely, a distribution of order 0. Our object now is to construct an extension of the domain of definition of the Stieltjes integral defined by the integrator ν, and we shall do so by making use of the additional information (given in (8.5) above) that the functional is a positive one; that is, that $\mu(f) \geqslant 0$ for every f in C_0 such that $f(t) \geqslant 0$ for all t.

8.2.2 Let μ be any linear functional on C_0 defined by a Stieltjes integral as in (8.7). We shall denote by $\bar{C}_0$ the set of all functions which are the limits of monotone increasing sequences of functions in C_0. That is to say, a function g belongs to $\bar{C}_0$ if and only if there exists a sequence (f_n) of functions in C_0 such that

(i) $f_n(t) \leqslant f_{n+1}(t)$ for all t and every n, and
(ii) $g(t) = \lim_{n \to \infty} f_n(t)$ for all t.

We can extend μ as a functional on $\bar{C}_0$ by monotone increasing limits as follows.

If g belongs to $\bar{C}_0$ let (f_n) be any sequence in C_0 which is monotone increasing with limit g and write

$$\mu(g) \equiv \lim_{n \to \infty} \mu(f_n) . \qquad (8.8)$$

It may be shown that this defines $\mu(g)$ uniquely, no matter which particular sequence (f_n) converging to g is chosen. Further, it is clear that if g is a function which belongs to C_0 itself then g is necessarily a member of $\bar{C}_0$ (we need only consider the monotone increasing sequence obtained by taking $f_n = g$ for every n). The value of $\mu(g)$ given by (8.8) is the same as that originally assigned to it by its definition as a functional on C_0. That is, the extension of μ by means of (8.8) is consistent. In general, of course, the members of $\bar{C}_0$ will not belong to C_0. Indeed an arbitrary function g in $\bar{C}_0$ need not vanish outside a finite interval, need not be continuous everywhere, and may even assume infinite values. In the same way we must accept the possibility that if g belongs to $\bar{C}_0$ then $\mu(g)$ may turn out to have the value $+\infty$ (though never $-\infty$).

Dually we denote by $\underline{C}_0$ the set of all functions h which are the limits of monotone decreasing sequences of functions in C_0. The functional μ can be extended to $\underline{C}_0$ by means of monotone decreasing limits in the sense that if h is any function in $\underline{C}_0$ and if (f_n) is a sequence of functions in C_0 which is monotone decreasing with limit h, then we can define $\mu(h)$ uniquely as $\lim_{n \to \infty} \mu(f_n)$. This time we note that $\mu(h)$ may take the value $-\infty$ (though never the value $+\infty$).

We are now in a position to advance a general definition of the extended functional μ or, what comes to the same thing, of the extended Stieltjes integral which lies behind it. We do this by defining new upper and lower integrals for an arbitrary function f in terms of the members of the function classes $\bar{C}_0$ and $\underline{C}_0$ defined above.

For an arbitrary function f we write,

$$\mu^*(f) \equiv \int^* f \mathrm{d}\mu = \inf_g \mu(g) \quad \text{where } g \text{ in } \bar{C}_0 \text{ and } g \geqslant f,$$

$$\mu_*(f) \equiv \int_* f \mathrm{d}\mu = \sup_h \mu(h) \quad \text{where } h \text{ in } \underline{C}_0 \text{ and } h \leqslant f.$$

The following properties are easy consequences of the definitions.

(i) $\mu_*(-f) = -\mu^*(f)$
(ii) $\mu^*(f + g) \leqslant \mu^*(f) + \mu^*(g)$
(iii) $\mu_*(f + g) \geqslant \mu_*(f) + \mu_*(g)$.

We say that f is **widely integrable** for μ if and only if the numbers $\mu^*(f)$ and $\mu_*(f)$ are equal. If, in addition, their common value is finite then we say that f is **integrable** for μ and write

$$\mu_*(f) = \mu^*(f) \equiv \mu(f) \equiv \int f \mathrm{d}\mu \tag{8.9}$$

8.2.3 *Remark.* The fact that we are now accepting that functions may take infinite values does cause certain technical difficulties in dealing, for example, with pointwise sums such as $f + g$. We can cope with this situation by adding

$+\infty$ and $-\infty$ as ideal elements to the real number system and adopting the following conventions for the arithmetic of this extended number system:

$$(+\infty.x) = +\infty \quad (0 < x \leqslant +\infty)$$
$$(+\infty.x) = -\infty \quad (-\infty \leqslant x < 0)$$
$$(-\infty.x) = -\infty \quad (0 < x \leqslant +\infty)$$
$$(-\infty.x) = +\infty \quad (-\infty \leqslant x < 0)$$
$$x/+\infty = x/-\infty = 0 \quad (-\infty < x < +\infty)$$
$$+\infty + x = +\infty \quad (x > -\infty)$$
$$-\infty + x = -\infty \quad (x < +\infty)$$
$$-\infty.0 = +\infty.0 = 0 .$$

(Combinations like $+\infty + (-\infty)$, $+\infty/+\infty$, etc. are left undefined).

Alternatively, we may tacitly agree to perform arithmetic operations on functions only at those points at which the functions concerned take finite values, leaving combinations like $f + g$ undefined at some points. In the event this often turns out to be unimportant since, in a sense to be explained below, the set of all such points may be described as "negligible" with respect to the functional μ in question.

8.2.4 If we denote by $\mathcal{L}^1(\mu)$ the set of all real, finite-valued functions which are integrable with respect to a given positive linear functional μ, originally defined on C_0, then the following results may be established:

(i) $\mathcal{L}^1(\mu)$ is a linear space, and μ is a positive linear functional on $\mathcal{L}^1(\mu)$.
(ii) f is integrable for μ if and only if $|f|$ is integrable for μ.
(iii) μ is continuous under monotone limits on $\mathcal{L}^1(\mu)$. That is, if the sequence (f_n) of functions in $\mathcal{L}^1(\mu)$ converges monotonely to f, then f is at least widely integrable for μ, and

$$\mu(f) = \lim_{n \to \infty} \mu(f_n) .$$

Apart from this the extension process described above effectively defines μ as a **set function** on a certain family of sets of real numbers. Given any set A of real numbers, we say that the **characteristic function** of A is the function ϕ_A defined by

$$\phi_A(t) = \begin{cases} 1 & \text{for all } t \text{ in } A \\ 0 & \text{for all } t \text{ not in } A . \end{cases}$$

If the characteristic function of A is an integrable function for μ, then A is said to be an integrable set with respect to μ, and we write

$$\mu(A) \equiv \mu(\phi_A) = \int \phi_A \, d\mu . \tag{8.10}$$

The number $\mu(A)$ is called the μ-**measure** of the set A; if $\mu(A) = 0$ then A is said to be a **set of measure zero** for μ or a μ-**negligible** set. A property P which holds everywhere, except possibly at the points of a set which is of measure zero for μ, is said to hold **almost everywhere** for μ. In many applications of integration theory this concept of measure zero is all that is really needed. For completeness, however, we give a general definition of measureability.

A bounded, closed, set of real numbers is said to be a **compact** set. It can be shown that every compact set K is an integrable set with respect to every (extended) positive linear functional μ. Accordingly, an arbitrary set A of real numbers is said to be **measureable** for μ if and only if the intersection $A \cap K$ is an integrable set for every compact K. (Roughly speaking the sets measurable for μ are those which are "locally integrable" sets for μ.)

8.3 LEBESGUE AND RIEMANN INTEGRALS

8.3.1 In what follows we confine attention to the particular case of the positive linear functional m defined on C_0 by the ordinary Riemann integral. The resulting extended functional is called the **Lebesgue integral**, and the associated set function $m(A)$ is called **Lebesgue measure**. (Note, incidentally, that $m([a,b]) = b - a$, the usual concept of "length" for an interval.) For brevity we shall often write

$$\int f dm = \int_{-\infty}^{+\infty} f(t) dt \equiv \int f . \qquad (8.11)$$

The Lebesgue integral, $\int f$, as defined above, is not confined to bounded integrands and is taken over the whole range from $-\infty$ to $+\infty$. Integration over a finite range $[a,b]$ can be treated within the context of the same general theory by simply noting that:

$$\int_{a}^{b} f(t) dt = \int_{-\infty}^{+\infty} f(t) \phi_{[a,b]}(t) dt \equiv \int f \phi_{[a,b]}$$

where $\phi_{[a,b]}$ denotes the characteristic function of the interval $[a,b]$.

By a **step function** over $[a,b]$ we shall mean any function s with the following properties:

(a) s vanishes identically outside $[a,b]$,
(b) there exists a finite set of points (t_k), $1 \leq k \leq n$, such that

$$a = t_0 < t_1 < t_2 < \ldots < t_n = b$$

and, $s(t) = \alpha_k$ (a finite constant) for $t_{k-1} < t < t_k$.

As is easy to verify, a step-function $\underline{s}$ belongs to the class $\underline{C_0}$ if and only if at each point of discontinuity, t_k, we have

$$\underline{s}(t_k) \geqslant \max\,(\alpha_k, \alpha_{k+1})\,.$$

Similarly, a step-function $\bar{s}$ belongs to the class $\bar{C_0}$ if and only if

$$\bar{s}(t_k) \leqslant \min\,(\alpha_k, \alpha_{k+1})\,.$$

Every step-function s has a well-defined integral over $[a,b]$ which is independent of the values which it assumes at its points of discontinuity.

Now the Riemann integral of a bounded function f over a finite interval $[a,b]$ can be defined in terms of the integrals of step-functions over $[a,b]$. (Without loss of generality we may assume that f vanishes identically outside $[a,b]$.) First we define the upper and lower Riemann integrals of f over $[a,b]$ as follows:

$$R\overline{\int_a^b} f(t)\mathrm{d}t = \inf_{\bar{s}(t)\leqslant f(t)} \int_a^b \bar{s}(t)\mathrm{d}t$$

$$R\underline{\int_a^b} f(t)\mathrm{d}t = \sup_{\underline{s}(t)\leqslant f(t)} \int_a^b \underline{s}(t)\mathrm{d}t$$

It will be noted that in defining the lower integral we use step functions belonging to $\underline{C_0}$ whereas for the upper integral we use step functions belonging the the class $\bar{C_0}$. It is easily confirmed that this restriction in no way affects the values of the upper and lower Riemann integrals. Moreover, it follows at once that

$$R\underline{\int_a^b} f(t)\mathrm{d}t \leqslant \int_* f \leqslant \int^* f \leqslant R\overline{\int_a^b} f(t)\mathrm{d}t$$

Since f is Riemann-integrable over $[a,b]$ if and only if its upper and lower integrals have the same value, we have the following result:

Let f be a bounded function which is integrable in the Riemann sense over the interval $[a,b]$. Then f is also integrable in the Lebesgue sense over that interval, and the Riemann and Lebesgue integrals have the same value. Hence, without ambiguity, we can write:

$$\int_a^b f(t)\mathrm{d}t\,,$$

without reference to the sense in which the integral is defined.

8.3.2 The decomposition of an arbitrary function into positive and negative components (Sec. 8.2.1) continues to hold, with minor modifications, even if f is allowed to take infinite values. We have

$$f(t) = f^+(t) - f^-(t) \quad \text{at all points at which } f(t) \text{ is finite,}$$

$$|f(t)| = f^+(t) + f^-(t) \quad \text{everywhere.}$$

From the definition of the Lebesgue integral given above it follows that if f is Lebesgue integrable then so also are the functions f^+ and f^-. Hence the Lebesgue theory yields an *absolute* integral in the sense that the integrability of f in the Lebesgue sense always implies the integrability of the function $|f|$. This shows at once that an *improper Riemann integral which is only conditionally convergent does not exist in the Lebesgue sense.*

Suppose instead that the (bounded) function f has an absolutely convergent improper Riemann integral from $-\infty$ to $+\infty$. Without loss of generality we may assume that $f(t) \geqslant 0$ everywhere. For $n = 1, 2, \ldots$ define

$$f_n(t) = f(t)\phi_{[-n,n]}(t) = \begin{cases} f(t), & \text{for } -n \leqslant t \leqslant n \\ 0, & \text{otherwise.} \end{cases}$$

Then,

$$R \int_{-\infty}^{+\infty} f(t)\,dt = \lim_{n \to \infty} R \int_{-n}^{n} f(t)\,dt = \lim_{n \to \infty} \int f_n = \int f .$$

A similar argument can be applied to absolutely convergent improper Riemann integrals of the second kind. Hence we have the result:

If an improper Riemann integral (of the first, second, or third kind) is absolutely convergent then the corresponding Lebesgue integral also exists and has the same value.

8.4 RIEMANN-STIELTJES AND LEBESGUE-STIELTJES INTEGRALS

8.4.1 The ostensible reasons for carrying out the programme of extending the range of definition of positive linear functionals on C_0 were the limitations of the elementary Riemann-type integration theory as illustrated in Sec. 8.1.3. In so far as the ordinary Riemann integral is concerned, it is very simple to show how the Lebesgue theory disposes of the difficulty quoted. The example given was of a sequence of functions (f_m) where

$$f_m(t) = \lim_{n \to \infty} [\cos(m!\pi t)]^{2n} \quad \text{and} \quad \int_0^1 f_m(t)\,dt = 0 .$$

In view of the discussion in Sec. 8.3 there is no need to state in which sense the integrals are to be understood, since the Riemann and Lebesgue integrals for the functions f_m are identical. For the limit function f we have seen that the Riemann integral does not exist. However, we have only to note that if $f_m(t) = 1$ for some m, then $t = p/q$ where $m \geqslant q$; thus, if $n \geqslant m$ we would have $n \geqslant q$ and so $f_n(t) = 1$. Hence, by the fact that the Lebesgue integral is continuous under monotone limits,

$$\int_0^1 f(t)\,dt = \lim_{m \to \infty} \int_0^1 f_m(t)\,dt = 0.$$

Let us turn instead to the problem posed by Stieltjes integrals of the form

$$\int_{-\infty}^{+\infty} g(t)\,du_c(t)$$

where g is not necessarily continuous at the origin. To begin with, recall that from the elementary Riemann–Stieltjes theory we know that the functional δ is defined on C_0 by

$$\delta(f) = \int_{-\infty}^{+\infty} f(t)\,du_c(t) = f(0)$$

and that this result is quite independent of the number c (see Sec. 2.4). Now take $g = u_0$; this is the characteristic function of the open interval $(0, +\infty)$ and can readily be expressed as the limit of a monotone increasing sequence (f_m) of functions in C_0 each such that $\delta(f_m) = 0$. (See, for example, Fig. 8.1)

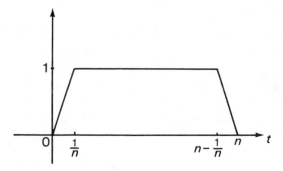

Fig. 8.1.

Hence we must have

$$\int_{-\infty}^{+\infty} u_0(t)\,\mathrm{d}u_c(t) = \lim_{m \to \infty} \int_{-\infty}^{+\infty} f_m(t)\,\mathrm{d}u_c(t) = 0$$

again independently of c.

Similarly we can show that

$$\int_{-\infty}^{+\infty} u_1(t)\,\mathrm{d}u_c(t) = 1 ,$$

and the way is open to development of an extended theory of the delta function as a Lebesgue–Stieltjes integral, with a virtually unrestricted sampling property.

Bibliographical Notes

Of the specialised texts on generalised functions currently available the multi-volume work by Gel'fand, Shilov, and others is almost certainly the most comprehensive and generally useful. For most practical applications the general reader is likely to find his needs satisfied by Chapters I and II of the first volume of the series, although it should perhaps be made clear that these two chapters comprise some two hundred pages! The authors themselves claim that these chapters "contain the standard minimum which must be known by all mathematicians and physicists who have to deal with generalised functions". In fact there is considerably more material than the term "standard minimum" might lead one to expect. The mode of treatment adopted is essentially the linear functional approach associated with Schwartz and outlined in Chapter 7 of the present book. It is hoped that the account given there will serve as a useful introduction to the Gel'fand and Shilov text and make the material of their Volume I more readily accessible.

Much the same ground is covered in the book by Zemanian. This is generally rather easier to follow than Gel'fand and Shilov and is liberally supplied with exercises. On the other hand, unlike the Russian text, it is almost exclusively concerned with the case of a single real variable. This is in itself by no means unreasonable since applications dealt with by Zemanian are primarily in the field of linear systems theory, and the book does contain a most useful introductory account of the theory of passive systems. Once again it is the linear functional approach of Schwartz which is used throughout, and our Chapter 7 should be of use as an introduction.

The book by Jones differs from both the preceding texts in so far as the actual definition and treatment of generalised functions are concerned. A generalised function is defined as an equivalence class of certain sequences of ordinary functions and, as a result, the reader is obliged to become familiar with a new terminology (though admittedly a very natural and straightforward one). It is a comprehensive text and is very well supplied with exercises, but the general reader is likely to find it somewhat heavy going to begin with. A most useful introduction to the terminology and to the general point of view of the book is to be found in the earlier, short treatise of Sir James Lighthill, of which indeed the fuller treatment in Jones is a natural development.

For those interested in the historical aspects of the subject the 'Operational Calculus' of van der Pol and Bremmer contains much information, particularly on the delta function itself. And in this connection it is a most interesting and rewarding exercise to consult Dirac's original remarks about $\delta(t)$ in his 'Quantum Mechanics' — arguably still one of the best and most readable heuristic introductory accounts. The definition of the delta function as a Stieltjes integral is also discussed in van der Pol and Bremmer but has been given comparatively little attention in the published literature since. An exception to this, of considerable intrinsic interest and importance, is the treatment of linear systems theory given in the 'Mathematics of Dynamical Systems' of Rosenbrock and Storey. In this a careful and thorough account of the elementary Riemann-Stieltjes integral is used as the basis of a restricted but an entirely rigorous theory of the delta function and of impulse response. Finally, when it comes to the theory of distributions proper, the two volumes of Laurent Schwartz are of course crucially important both from a historical point of view and as the prime source and reference for most of all the subsequent developments. At the same time it should be said that this is by no means an easy book to read, even for the specialist mathematician, and it is certainly not recommended as a suitable text for the beginner.

We do not attempt here to list even a selection of the many books on Functional Analysis in which applications of generalised functions are to be found. Most of them will contain a brief account of the theory of distributions, but sooner or later a reference to one or other of the specialised treatises referred to above will be found to be necessary. Attention should, however, be drawn to the book on 'Applied Functional Analysis' by David Griffel, currently in preparation, which should form a useful companion text to the present volume.

References

P. A. M. DIRAC — The Principles of Quantum Mechanics.
   Oxford University Press, 1947.

I. M. GEL'FAND, G. E. SHILOV, et al. — Generalised Functions.
   Vol. 1 Properties and Operations
   Vol. 2 Spaces of Fundamental and Generalised Functions
   Vol. 3 Theory of Differential Equations
   Vol. 4 Applications of Harmonic Analysis
   Vol. 5 Integral Geometry and Representation Theory
   Academic Press, 1964–1966.

D. GRIFFEL — Applied Functional Analysis.
   Ellis Horwood (in preparation).

D. S. JONES — Generalised Functions.
   McGraw-Hill, 1966.

M. J. LIGHTHILL — An Introduction to Fourier Analysis and Generalised
   Functions.
   Cambridge University Press, 1960.

H. H. ROSENBROCK and C. STOREY — Mathematics of Dynamical Systems.
   Thomas Nelson, 1970.

L. SCHWARTZ — Théorie des Distributions.
   Herman, 1957–1959.

van der POL and BREMMER — Operational Calculus, based on the two-sided
   Laplace Integral.
   Cambridge University Press, 1960.

A. H. ZEMANIAN — Distribution Theory and Transform Analysis.
   McGraw-Hill, 1965.

References

P. A. M. DIRAC — *The Principles of Quantum Mechanics*.
Oxford University Press, 1958.

Solutions to Exercises

CHAPTER 2

Exercises I

1. (a) $u\{(t-a)(t-b)\} = \begin{cases} 1 \text{ for } t < \min(a,b) \text{ and } t > \max(a,b) \\ 0 \text{ for } \min(a,b) < t < \max(a,b); \end{cases}$

(b), (c) $u(e^t - \pi) = u(t - \log \pi) = 1 \text{ for } t > \log \pi, = 0 \text{ for } t < \log \pi$;

(d) $u(\sin t) = \begin{cases} 1 \text{ for } 2n\pi < t < (2n+1)\pi \\ 0 \text{ for } (2n-1)\pi < t < 2n\pi \end{cases}, n = 0, \pm 1, \pm 2, \dots$

(e) $u(\cos t) = \begin{cases} 1 \text{ for } (2n - \frac{1}{2})\pi < t < (2n + \frac{1}{2})\pi \\ 0 \text{ for } (2n - \frac{3}{2})\pi < t < (2n - \frac{1}{2})\pi \end{cases}, n = 0, \pm 1, \pm 2, \dots$

(f) $u(\sinh t) = u(t)$; (g) $u(\cosh t) = 1$, for all t.

2. (a) $\operatorname{sgn}(t^2 - 1) = \begin{cases} +1 \text{ for } |t| > 1 \\ -1 \text{ for } |t| < 1 \end{cases}$; (b) $\operatorname{sgn}(e^{-t}) = 1$, for all t ;

(c) $\operatorname{sgn}(\tan t) = \begin{cases} +1 \text{ for } n\pi < t < (2n+1)\pi/2 \\ -1 \text{ for } (2n-1)\pi/2 < t < n\pi \end{cases} n = 0, \pm 1, \pm 2, \dots$

(d) $\operatorname{sgn}(\sin 1/t) = \begin{cases} +1 \text{ for } 1/(2n+1)\pi < t < 1/2n\pi \\ -1 \text{ for } 1/2n\pi < t < 1/2(n-1)\pi \end{cases}, n = 1, 2, 3, \dots$

Also, $\operatorname{sgn}(\sin 1/t) = +1$ for $t > 1/\pi$, and to complete the definition for negative values of t we need only note that the function is *odd*.

(e) $t^2 \operatorname{sgn} t = \begin{cases} t^2 \text{ for } t > 0 \\ -t^2 \text{ for } t < 0 \end{cases}$; (f) $\{\operatorname{sgn}(\sin t)\} \sin t = |\sin t|$;

(g) $\sin t\{\text{sgn}(\cos t)\} = \left[\begin{array}{l} \sin t \text{ for } (4n-1)\dfrac{\pi}{2} < t < (4n+1)\dfrac{\pi}{2} \\[2mm] -\sin t \text{ for } (4n+1)\dfrac{\pi}{2} < t < (4n+3)\dfrac{\pi}{2} \end{array}\right]$

where $n = 0, \pm1, \pm2, \ldots$;

3. Let $g(t) = tu_0(t)$; then $g(b) - g(a) = \displaystyle\int_a^b u(t)\,dt$.

4. (a) $-3t^2\text{sgn}\, t$;   (b) $e^{|t|}\text{sgn}\, t$;   (c) $-e^{-|t|}\,\text{sgn}\, t$;   (d) $\text{sgn}\, t.\cos t$;

    (e) $\{\text{sgn}(\sin t)\}\cos t$;   (f) $[\text{sgn}(\sin t)]\cos t$;   (g) $\text{sgn}\, t.\cosh t$.

5. (a) $f(t_0 + h) - f(t_0) = h\{f'(t_0) + \epsilon\}$ where $f'(t_0)$ is finite and $\epsilon \to 0$ with h ;

    (b) For $t \neq 0$ we have $f'(t) = 2t \sin 1/t - \cos 1/t$.

        For $t = 0$, $f'(0) = \displaystyle\lim_{h\to0} \frac{h^2 \sin 1/h}{h} = \lim_{h\to0} h \sin 1/h = 0$.

        Neither $f'(0+)$ nor $f'(0-)$ exist ;

    (c) For $t \neq 0$ we have $g'(t) = 2t \sin 1/t^2 - \dfrac{2}{t}\cos 1/t^2$.

        For $t = 0$, $g'(0) = \displaystyle\lim_{h\to0} \frac{h^2 \sin 1/h^2}{h} = \lim_{h\to0} h \sin 1/h^2 = 0$.

6. (a) $\displaystyle\lim_{n\to\infty} \left[\frac{1}{2} + \frac{1}{\pi}\tan^{-1} nt\right] = \left[\begin{array}{l} 1 \text{ for } t > 0 \\ 1/2 \text{ for } t = 0 \\ 0 \text{ for } t < 0 \end{array}\right] = u_{1/2}(t);$

    (b) $e^{-x} \to 0$ as $x \to +\infty$ and $\to +\infty$ as $x \to -\infty$.
    Hence $\exp(-e^{-x}) \to 1$ as $x \to +\infty$ and $\to 0$ as $x \to -\infty$, so that

    $\lim \exp(-e^{-nt}) = \left[\begin{array}{l} 1 \text{ for } t > 0 \\ 1/e \text{ for } t = 0 \\ 0 \text{ for } t < 0 \end{array}\right] \equiv u_{1/e}(t)$.

Exercises II

1. $\displaystyle\int_{-\infty}^{+\infty} f(t)\delta(-t)\,dt = \int_{-\infty}^{+\infty} f(-\tau)\delta(\tau)\,d\tau = \Big[f(-\tau)\Big]_{\tau=0} = f(0)$.

2. $\displaystyle\int_{-\infty}^{+\infty} f(t)\left[\frac{u(t-a+\eta) - u(t-a)}{\eta}\right]dt = \frac{1}{\eta}\int_{a-\eta}^{a} f(t)\,dt = f(\xi)$, for some ξ

    such that $a - \eta < \xi < a$.

3. A formal integration by parts shows that, at least for any f continuously differentiable on a neighbourhood of the origin,

$$\int_{-\infty}^{+\infty} f(t)\{u(-t)\}'dt = -f(0).$$

However, since $u(-t) = 1 - u(t)$, it is immediately obvious that

$$\{u(-t)\}' = -\delta(t).$$

Similarly, since sgn $t = u(t) - u(-t)$, we can at once deduce that

$$\int_{-\infty}^{+\infty} f(t)\{\text{sgn } t\}'dt = 2f(0).$$

Exercises III

1. $$\int_{-\infty}^{+\infty} \cos t \, d_n(t)dt = 2n \int_0^{1/2n} \cos t \, dt = 2n.\sin 1/2n \to 1 \text{ as } n \to \infty.$$

2. $$\int_{-\infty}^{+\infty} \cos t \, g_n(t)dt =$$

$$2\left[n \int_0^{1/n} \cos t \, dt - n^2 \int_0^{1/n} t \cos t \, dt \right] = 2n^2 \left(1 - \cos \frac{1}{n} \right) \to 1.$$

3. $$\int_{-\infty}^{+\infty} \cos t \, h_n(t)dt =$$

$$2\left[2n^2 \int_0^{1/2n} t\cos t \, dt + 2n \int_{1/2n}^{1/n} \cos t \, dt - 2n^2 \int_{1/2n}^{1/n} t\cos t \, dt \right]$$

$$= 2\left[4n^2\cos 1/2n - 2n^2\{1 + \cos 1/n\} \right].$$

(That this tends to 1 as $n \to \infty$ is most easily seen by using Maclaurin expansions for the cosine terms; similarly for the limit in (2) above). Note finally that, since $h_n(0) = 0$ for each n, we must have $\lim_{n \to \infty} h_n(0) = 0$.

4. (a) For every partition of $[a,b]$ we would have

$$\sum_{r=1}^{n} f(t_r)\Delta_r\phi = f(t_1)\{u_c(t_1 - a) - u_c(0)\} = f(t_1)(1 - c)$$

and hence $\int_a^b f(\tau)du_c(\tau - a) = f(a)(1 - c)$;

The argument for (b) is similar.

CHAPTER 3

Exercises I

1. (a) $\cos t$; (b) $\sin t + \delta(t)$; (c) $1 + 2e\,\delta(t-1)$.

2. (a) 5 ; (b) $1/3$; (c) e^{-4} ; (d) $\sinh 4$; (e) $\pi + 1/2e^2$;

   (f) $\displaystyle\sum_{k=1}^{n} k^2 = \frac{1}{6}\,[n(n+1)(2n+1)]$; (g) $\dfrac{1}{2} - 1 + \dfrac{1}{2} = 0$; (h) $2e^{\pi}$.

Exercises II

1. (a) $\delta(t+1) - \delta(t-1)$; (b) $\{u(t) - 1\}\sin t - \delta(t)$;

   (c) $\{u(t - \pi/2) - u(t - 3\pi/2)\}\cos t + \delta(t - \pi/2) + \delta(t - 3\pi/2)$;

   (d) $e^4\,\delta(t+2) + \delta(t) - e^{-4}\,\delta(t-2) - 2e^{-2t}\{u(t+2) + u(t) - u(t-2)\}$.

2. (a) $\operatorname{sgn} t, 2\,\delta(t)$; (b) $-e^{-|t|}\operatorname{sgn} t, e^{-|t|} - 2\,\delta(t)$;

   (c) $\operatorname{sgn} t.\cos |t|, 2\,\delta(t) - \operatorname{sgn} t.\sin t$.

3. $\displaystyle\lim_{t\to 0+} f(t) = \lim_{x\to +\infty}\left[\frac{e^x - e^{-x}}{e^x + e^{-x}}\right] = +1$

   $\displaystyle\lim_{t\to 0-} f(t) = \lim_{x\to -\infty}\left[\frac{e^x - e^{-x}}{e^x + e^{-x}}\right] = -1$

   $f'(t) = -1/[t^2 \cosh^2(1/t)]$, $\mathcal{D}f(t) = 2\,\delta(t) - 1/[t^2\cosh^2(1/t)]$.
   Finally, note that $\displaystyle\lim_{t\to 0} 1/[t^2\cosh^2(1/t)] = \lim_{x\to\infty} x^2/\cosh^2 x = 0$.

Exercises III

1. (a) $-b$, $2ab$, $b^3 - 3a^2 b$; (b) 1 ; (c) 19 .

2. $\displaystyle\int_{-\infty}^{+\infty} f(t)\delta'(-t)\,\mathrm{d}t = \int_{-\infty}^{+\infty} f(-\tau)\delta'(\tau)\,\mathrm{d}\tau = \left[\frac{-\mathrm{d}}{\mathrm{d}\tau}f(-\tau)\right]_{\tau=0} = +f'(0)$.

3. $\phi(t)\delta'(t) = \phi(0)\delta'(t) - \phi'(0)\delta(t)$, and so

   $\dfrac{\mathrm{d}}{\mathrm{d}t}\{\phi(t)\delta'(t)\} = \phi(0)\delta''(t) - \phi'(0)\delta'(t)$.

   Also, $\phi'(t)\delta'(t) + \phi(t)\delta''(t) = [\phi'(0)\delta'(t) - \phi''(0)\delta(t)]$
   $\qquad\qquad\qquad\qquad\qquad + [\phi(0)\delta''(t) - 2\phi'(0)\delta'(t) + \phi''(0)\delta(t)]$.

Exercises IV

1. (a) $\int_{-1}^{0} \sinh 2t . \delta(5t+2)\,dt = \int_{-3}^{2} \sinh 2\left(\frac{x-2}{5}\right)\delta(x)\frac{dx}{5} = \frac{-1}{5}\sinh\frac{4}{5}$;

(b) $\int_{-2\pi}^{2\pi} e^{\pi t}\,\delta(t^2 - \pi^2)\,dt = \frac{1}{2\pi}\int_{-2\pi}^{2\pi} e^{\pi t}\{\delta(t+\pi)+\delta(t-\pi)\}\,dt = \frac{\cosh \pi^2}{\pi}$.

(c) For $-\pi < \theta < 0$, $\cos\theta$ increases from -1 to $+1$, with $\cos(-\pi/2)=0$; hence $u(\cos\theta) = u(\theta + \pi/2)$ for this range of values of θ. Similarly, for $0 < \theta < \pi$, $\cos\theta$ decreases from $+1$ to -1, with $\cos(\pi/2)=0$; hence $u(\cos\theta) = -u(\theta - \pi/2)$ for this range of values of θ. Thus, for $-\pi < \theta < \pi$,

$$\delta(\cos\theta) = \frac{1}{(-\sin\theta)}\,\delta(\theta + \pi/2) - \frac{1}{(-\sin\theta)}\,\delta(\theta - \pi/2) .$$

Hence

$$\int_{-\pi}^{\pi} \cosh\theta . \delta(\cos\theta)\,d\theta = \int_{-\pi}^{\pi} \cosh\theta . \{\delta(\theta+\pi/2)+\delta(\theta-\pi/2)\}\,d\theta$$

$$= 2\cosh \pi/2 .$$

(d) $u(\sin\pi t) = \displaystyle\sum_{m=-\infty}^{+\infty} [u(t-2m) - u\{t-(2m+1)\}]$

Hence,

$$\frac{d}{dt}u(\sin\pi t) = \sum_{m=-\infty}^{+\infty} [\delta(t-2m) - \delta\{t-(2m+1)\}]$$

But,

$$\delta(\sin\pi t) = \frac{du(\sin\pi t)}{d(\sin\pi t)} = \frac{du(\sin\pi t)}{dt}\Big/\frac{d}{dt}(\sin\pi t)$$

$$= \frac{1}{\pi\cos\pi t}\sum_{m=-\infty}^{+\infty} [\delta(t-2m) - \delta\{t-(2m+1)\}]$$

$$= \sum_{m=-\infty}^{+\infty}\left[\frac{\delta(t-2m)}{\pi\cos 2m\pi} - \frac{\delta\{t-(2m+1)\}}{\pi\cos(2m+1)\pi}\right] = \frac{1}{\pi}\sum_{m=-\infty}^{+\infty} \delta(t-m),$$

so that

$$\int_{-\infty}^{+\infty} e^{-|t|}\delta(\sin\pi t)\,dt = \frac{1}{\pi}\left[1 + 2\sum_{m=1}^{\infty} e^{-m}\right] = \frac{1}{\pi}\left[1 + \frac{2}{e-1}\right] = \frac{e+1}{\pi(e-1)} .$$

2.   (a) $u(\sin |t|) = \sum\limits_{n=1}^{\infty} (-1)^n u(t - n\pi) + \sum\limits_{n=1}^{\infty} (-1)^{n+1} u(t + n\pi)$.

$\dfrac{d}{dt} \sin |t| = \operatorname{sgn} t.\cos t = (-1)^n$ for $t = n\pi$, $= (-1)^{n+1}$ for $t = -n\pi$,

where $n = 1, 2, 3, \ldots$

$$\dfrac{d}{dt} u(\sin |t|) = \sum\limits_{n=1}^{\infty} (-1)^n \,\delta(t - n\pi) + \sum\limits_{n=1}^{\infty} (-1)^{n+1} \,\delta(t + n\pi).$$

Hence,

$$\delta(\sin |t|) = \sum\limits_{n=1}^{\infty} \{\delta(t - n\pi) + \delta(t + n\pi)\}.$$

(b) $u(\cos \pi t/2) = \sum\limits_{m=-\infty}^{+\infty} (-1)^{m+1} \delta\{t - (2m + 1)\}$

$$\dfrac{d}{dt} u(\cos \pi t/2) = \sum\limits_{m=-\infty}^{+\infty} (-1)^{m+1} \,\delta\{t - (2m + 1)\}.$$

$\dfrac{d}{dt} (\cos \pi t/2) = \dfrac{-\pi}{2} \sin \pi t/2 = -\pi/2$ when $t = 2m + 1$ (m even) and

$= +\pi/2$ when $t = 2m + 1$ (m odd). It follows that

$$\delta(\cos \pi t/2) = \dfrac{1}{\dfrac{-\pi}{2} \sin \pi t/2} \sum\limits_{m=-\infty}^{+\infty} (-1)^{m+1} \delta\{t - (2m + 1)\}$$

$$= \dfrac{2}{\pi} \sum\limits_{-\infty}^{+\infty} \delta\{t - (2m + 1)\}.$$

(c) $u(e^t) \equiv 1$. Hence $\dfrac{d}{dt} u(e^t) = 0$, and so $\delta(e^t) = 0$.

(d) $\delta'(\theta^2 - \pi^2) = \dfrac{d}{d\theta}\delta(\theta^2 - \pi^2)\Big/\dfrac{d}{d\theta}(\theta^2 - \pi^2) = \left[\dfrac{1}{2\theta}\dfrac{1}{2\pi}\{\delta'_\pi(\theta) + \delta'_{-\pi}(\theta)\}\right]$

$$= \dfrac{1}{2\pi}\left[\dfrac{\delta'(\theta + \pi)}{2\theta} + \dfrac{\delta'(\theta - \pi)}{2\theta}\right]$$

$$= \dfrac{1}{2\pi}\left[\dfrac{\delta'(\theta - \pi)}{2\pi} + \dfrac{\delta(\theta - \pi)}{2\pi^2} - \dfrac{\delta'(\theta + \pi)}{2\pi^2} + \dfrac{\delta(\theta + \pi)}{2\pi^2}\right].$$

(e) $\delta'(\sinh 2x) = \dfrac{d}{dx}\delta(\sinh 2x)\Big/\dfrac{d}{dx}(\sinh 2x) = \dfrac{1}{2\cosh 2x}\left[\dfrac{\delta(x)}{2}\right] = \dfrac{1}{4}\delta(x)\,.$

3. From the hypothesis we have $u\{\phi(t)\} = u(t - c)$ so that

$$\delta\{\phi(t)\} = \dfrac{du\{\phi(t)\}}{d\{\phi(t)\}} = \dfrac{du\{\phi(t)\}}{dt}\Big/\dfrac{d\phi}{dt} = \dfrac{1}{\phi'(t)}\delta(t - c) = \dfrac{1}{\phi'(c)}\delta(t - c)\,.$$

$$\delta'\{\phi(t)\} = \dfrac{d\delta\{\phi(t)\}}{d\{\phi(t)\}} = \dfrac{d\delta\{\phi(t)\}}{dt}\Big/\dfrac{d\phi}{dt} = \dfrac{1}{\phi'(t)}\left[\dfrac{\delta'(t - c)}{\phi'(c)}\right]$$

$$= \dfrac{1}{\phi'(c)}\left[\dfrac{1}{\phi'(t)}\delta'(t - c)\right] = \dfrac{1}{\phi'(c)}\left[\dfrac{\delta'(t - c)}{\phi'(c)} + \dfrac{\phi''(c)}{[\phi'(c)]^2}\delta(t - c)\right].$$

Alternatively, for any continuously differentiable function f we have

$$\int_a^b f(t)\delta'\{\phi(t)\}\,dt = \int_{\phi(a)}^{\phi(b)} f\{\phi^{-1}(x)\}\delta'(x)\{\phi^{-1}(x)\}'\,dx$$

$$= -\dfrac{d}{dx}\Big[f\{\phi^{-1}(x)\}\{\phi^{-1}(x)\}'\Big]_{x=\phi(c)} = -\dfrac{d}{dx}\left[f(t)\dfrac{dt}{dx}\right]_{t=c}$$

$$= -\left[\dfrac{d}{dt}\left[\dfrac{f(t)}{\phi'(t)}\right]\Big/\dfrac{d\phi}{dt}\right]_{t=c}$$

$$= -\dfrac{1}{\phi'(c)}\left[\dfrac{f'(c)}{\phi'(c)} - \dfrac{f(c)\phi''(c)}{[\phi'(c)]^2}\right].$$

CHAPTER 4

Exercises I

1. (a) linear and time-invariant ;   (b) linear, but time-varying ;
    (c) linear, but time-varying ;   (d) non-linear, time-invariant ;
    (e) non-linear, time-invariant .

2. $T[x] + T[-x] = T[x - x] = T[0] = 0$; hence, $T[-x] = -T[x]$. Let $r = m/n$, where m and n are positive integers; if x is any input function, write $f \equiv x/n$.

    Then for an additive operator T we have

    $$T[mx] = mT[x] , \quad \text{and} \quad nT[f] = T[nf] = T[x] ,$$

    and the result follows.

3. Linearity is a trivial consequence of the linearity of the integral. Now the response to the translated input $x(t-a)$ is given by

    $$\int_{-\infty}^{+\infty} x(v-a)h(t,v)dv = \int_{-\infty}^{+\infty} x(\tau)h(t,\tau+a)d\tau .$$

    If the system is time-invariant, this must be equal to

    $$y(t-a) \equiv \int_{-\infty}^{+\infty} x(\tau)h(t-a,\tau)d\tau .$$

    Hence, $h(t,\tau+a) = h(t-a,\tau)$ so that for $\tau = 0$, we have $h(t,a) = h(t-a,0)$. Result follows if we write $h(t-\tau) \equiv h(t-\tau,0)$.

Exercises II

1. $\sigma(t) = \displaystyle\int_{-\infty}^{t} u(\tau)d\tau = \int_{0}^{t} d\tau = tu(t)$.

    $h(t) = \dfrac{d}{dt}\{tu(t)\} = t\delta(t) + 1u(t) = u(t)$.

2. (a) $u(t) * u(t) = \displaystyle\int_{-\infty}^{+\infty} u(t-\tau)u(\tau)d\tau = tu(t)$;

    (b) $e^{\alpha t} * \{u(t)\cos\omega t\} = e^{\alpha t}\displaystyle\int_{0}^{\infty} e^{-\alpha\tau}\cos\omega\tau \, d\tau = \dfrac{\alpha e^{\alpha t}}{\alpha^2 + \omega^2}$;

(c) $\{u(t) \sin t\} * \{u(t) \cos t\} = \displaystyle\int_0^t \sin(t - \tau) \cos \tau \, d\tau = \left[\dfrac{t}{2} \sin t\right] u(t)$;

(d) If $f(t) = p(t) * p(t)$, then:

$$f(t) = 0 \text{ for all } t < -1 \text{ , and for all } t > 1 \text{ ;}$$

for $-1 \leqslant t \leqslant 0$ we have $\displaystyle\int_{-1/2}^{t+1/2} d\tau = \left(t + \dfrac{1}{2}\right) - \left(-\dfrac{1}{2}\right) = t + 1$;

for $0 \leqslant t \leqslant +1$ we have $\displaystyle\int_{t-1/2}^{1/2} d\tau = \dfrac{1}{2} - \left(t - \dfrac{1}{2}\right) = 1 - t$.

Hence, $p(t) * p(t) = 1 - |t|$ for $|t| \leqslant 1$, and $= 0$ for $|t| > 1$.

3. $g_n'(t) \equiv \dfrac{d}{dt} \displaystyle\int_0^t \dfrac{(t - \tau)^n}{n!} f(\tau) d\tau = \int_0^t \dfrac{\partial}{\partial t}\left[\dfrac{(t - \tau)^n}{n!}\right] f(\tau)\, d\tau \ = g_{n-1}(t)$.

Since $g_n(0) = 0$, it follows that $g_n(t) = \displaystyle\int_0^t g_{n-1}(\tau) d\tau, n \geqslant 1$.

But $g_0(t) = \displaystyle\int_0^t f(\tau) d\tau$, and so g_n can be obtained from f by integrating $(n + 1)$ times from 0 to t.

4. $\displaystyle\int_0^t \dfrac{d\tau}{\sqrt{\tau}\sqrt{t - \tau}} = \int_0^{\sqrt{t}} \dfrac{2x\,dx}{x\sqrt{t - x}} = 2\left[\sin^{-1}(x/\sqrt{t})\right]_0^{\sqrt{t}} = \pi, \text{ for } t > 0$.

Hence, $g(t) * g(t) = u(t)$, and so for any continuous function f which vanishes for $t < 0$,

$$\{f(t) * g(t)\} * g(t) = f(t) * \{g(t) * g(t)\} = \int_0^t f(\tau) d\tau .$$

Exercises III

1. $\displaystyle\int_{-\infty}^{+\infty} [x^4 - 10x^2 + 1] [\delta_{\sqrt{2}}(x) * \delta_{\sqrt{3}}(x)] dx$

$= \displaystyle\int_{-\infty}^{+\infty} [x^4 - 10x^2 + 1] \delta_{(\sqrt{2}+\sqrt{3})}(x) dx$

which is zero, since $x = \sqrt{2} + \sqrt{3}$ is a root of $x^4 - 10x^2 + 1 = 0$.

2. (a) $e^{st} \to A e^{st}$; $H(s) = A$ for every s ;

   (b) $e^{st} \to e^{s(t-a)} = e^{-sa} e^{st}$; $H(s) = e^{-sa}$ for every s ;

   (c) $e^{st} \to s e^{st}$; $H(s) = s$ for every s ;

   (d) $e^{st} \to \int_{-\infty}^{t} e^{s\tau} \, d\tau = \frac{1}{s} e^{st}$; $H(s) = \frac{1}{s}$, $\operatorname{Re}(s) > 0$.

3. Impulse response of the system is given by

$$h(t) = \frac{d}{dt}\{u(t)\sin \alpha t\} = u(t)\alpha \cos \alpha t + \delta(t)\alpha \sin \alpha t = u(t)\alpha \cos \alpha t .$$

Hence, $y(t) = \alpha \int_{0}^{t} \sin \alpha(t-\tau) \cos \alpha\tau \, d\tau = \frac{\alpha t}{2} \sin \alpha t.$

CHAPTER 5

Exercises I

1. (a) $\int_{0}^{\infty} e^{-st} u(t) \, dt = \int_{0}^{\infty} e^{-st} \, dt = \frac{1}{s}$, provided $\operatorname{Re}(s) > 0$;

   (b) $\int_{0}^{\infty} e^{-st} [u(t) - u(t-1)] \, dt = \int_{0}^{1} e^{-st} \, dt = \frac{1 - e^{-s}}{s}$;

   (c) $\int_{0}^{\infty} e^{at} e^{-st} \, dt = \left[-\frac{1}{s-a} e^{-(s-a)t} \right]_{0}^{\infty} = \frac{1}{s-a}$, provided $\operatorname{Re}(s) > a$;

   (d) $\int_{0}^{\infty} e^{-st} t \, dt = \frac{1}{s^2}$, provided $\operatorname{Re}(s) > 0$. (Integrate by parts) ;

   (e) $\int_{0}^{\infty} e^{-st} \cosh bt \, dt = \frac{s}{s^2 - b^2}$, $\operatorname{Re}(s) > |b|$. (Use (c) and linearity) ;

   (f) $\int_{0}^{\infty} e^{-st} \sin \omega t \, dt = \frac{\omega}{s^2 + \omega^2}$, $\operatorname{Re}(s) > 0$. (Integrate by parts) ;

   (g) $\int_{0}^{\infty} e^{-st} \cos \omega t \, dt = \frac{s}{s^2 + \omega^2}$, $\operatorname{Re}(s) > 0$. (Integrate by parts).

2. $\mathcal{L}\{y''(t)\} = s^2 Y(s) - s y(0) - y'(0)$. Hence, taking transforms,

$$s^2 Y(s) - s + 2 + Y(s) = \frac{1}{s^2} \, ,$$

so that

$$Y(s) = \frac{1}{s^2(s^2 + 1)} + \frac{s - 2}{s^2 + 1} = \frac{1}{s^2} - \frac{1}{s^2 + 1} + \frac{s}{s^2 + 1} - \frac{2}{s^2 + 1}$$

$$= \frac{1}{s^2} + \frac{s}{s^2 + 1} - \frac{3}{s^2 + 1} . \text{ Hence, } y(t) = [t + \cos t - 3 \sin t] u(t) .$$

3. $Y(s) = \frac{s}{s^2 + 1} + \frac{2}{s^2 + 1} + \frac{1}{s^2 + 1} Y(s) .$

Hence, $Y(s) = \frac{s + 2}{s^2} = \frac{1}{s} + \frac{2}{s^2}$ and $y(t) = [1 + 2t] u(t) .$

4. $sX(s) - 8 = 2X(s) - 3Y(s)$

$sY(s) - 3 = Y(s) - 2X(s)$

giving $X(s) = \frac{5}{s + 1} + \frac{3}{s - 4}$ and $Y(s) = \frac{5}{s + 1} - \frac{2}{s - 4} .$

Hence, $x(t) = [5e^{-t} + 3e^{4t}] u(t), y(t) = [5e^{-t} - 2e^{4t}] u(t) .$

5. If $g(t) = u(t)/\sqrt{\pi t}$ then $g(t) * g(t) = u(t)$. (Ex. II, No. 4, of Chapter 4).

Hence $G(s).G(s) = 1/s$, whence $G(s) = \int_0^\infty e^{-st} \frac{1}{\sqrt{\pi t}} dt = \frac{1}{\sqrt{s}} \, .$

Exercises II

1. (a) $\frac{s^3 + 2}{s + 1} = s^2 - s + 1 + \frac{1}{s + 1}$; $f(t) = \delta''(t) - \delta'(t) + \delta(t) + e^{-t} u(t)$;

    (b) $\frac{\cosh s}{e^s} = \frac{e^s + e^{-s}}{2e^s} = \frac{1}{2}[1 + e^{-2s}]$; $f(t) = \frac{1}{2}[\delta(t) + \delta(t - 2)]$;

    (c) $\frac{1 - s^n e^{-ns}}{1 - se^{-s}} = \sum_{k=0}^{n-1} s^k e^{-ks}$; $f(t) = \sum_{k=0}^{n-1} \delta^{(k)}(t - k) .$

2. (a) $f(t) = u(t) - u(t-a)$; $F_0(s) = \dfrac{1}{s}[1 - e^{-as}]$;

$$\mathcal{L}\{f_T(t)\} = \frac{1}{s}\left[\frac{1 - e^{-as}}{1 - e^{-\pi s}}\right]$$

(b) $f(t) = \dfrac{1}{2}u(t) - u(t-a) + \dfrac{1}{2}u(t-\pi)$; $F_0(s) = \dfrac{1}{2s}[1 - 2e^{-as} + e^{-s\pi}]$

$$\mathcal{L}\{f_T(t)\} = \frac{1}{2s}\left[\frac{1 - 2e^{-as}e^{-s\pi}}{1 - e^{-\pi s}}\right].$$

3. $\mathcal{L}\{\delta(\cos t)\} = \mathcal{L}\left[\displaystyle\sum_{n=-\infty}^{+\infty}\delta\left[t - \left(n + \dfrac{1}{2}\right)\pi\right]\right]$

$$= \mathcal{L}\left[\sum_{n=0}^{\infty}\delta\left[t - \left(n + \frac{1}{2}\right)\pi\right]\right] = \sum_{n=0}^{\infty}e^{-(n+\frac{1}{2})\pi s} = e^{-\pi s/2}/[1 - e^{-\pi s}].$$

Exercises III

1. (a) $\mathcal{L}\{\cosh^2 at\} = \mathcal{L}\left[\dfrac{1 + \cosh 2at}{2}\right] = \dfrac{1}{2}\left[\dfrac{1}{s} + \dfrac{s}{s^2 - 4a^2}\right] = \dfrac{s^2 - 2a^2}{s(s^2 - 4a^2)}$

(b) $\mathcal{L}\{e^{-t}\sin^2 t\} = \dfrac{1}{2}\mathcal{L}\{e^{-t}(1 - \cos 2t)\} = \dfrac{1}{2}\left[\dfrac{1}{s+1} - \dfrac{s+1}{s^2 + 2s + 5}\right]$;

(c) $\mathcal{L}\{t^3 \cosh t\} = \mathcal{L}\left[\dfrac{t^3 e^t + t^3 e^{-t}}{2}\right] = \dfrac{3}{(s-1)^4} + \dfrac{3}{(s+1)^4}$;

(d) $\mathcal{L}\{t \cos at\} = -\dfrac{d}{ds}\left[\dfrac{s}{s^2 + a^2}\right] = -\dfrac{(s^2 + a^2) - 2s^2}{(s^2 + a^2)^2} = \dfrac{s^2 - a^2}{(s^2 + a^2)^2}$;

(e) $\mathcal{L}\{t^2 \sin at\} = +\dfrac{d^2}{ds^2}\left[\dfrac{a}{s^2 + a^2}\right] = \dfrac{6as^2 - 2a^3}{(s^2 + a^2)^3}$.

2. Put $g(t) \equiv f(t)/t$ so that $f(t) = tg(t)$, and $F(s) = -G'(s)$.

Then, $G(s) - \displaystyle\lim_{s\to\infty} G(s) = \int_s^{\infty} F(p)\,dp$, and $\displaystyle\lim_{s\to\infty} G(s) = 0$.

(a) $\mathcal{L}\left[\dfrac{\sin t}{t}\right] = \displaystyle\int_s^{\infty}\dfrac{dx}{x^2 + 1} = \int_0^{1/s}\dfrac{dy}{y^2 + 1} = \tan^{-1}(1/s)$;

(b) $\mathcal{L}\left[\dfrac{\cos at - \cos bt}{t}\right] = \displaystyle\int_0^{1/s}\left[\dfrac{x}{x^2+a^2} - \dfrac{x}{x^2+b^2}\right]dx = \dfrac{1}{2}\log\left[\dfrac{s^2+b^2}{s^2+a^2}\right] ;$

(c) $\mathcal{L}\left[\displaystyle\int_0^t \dfrac{\sin\tau}{\tau}\,d\tau\right] = \dfrac{1}{s}\tan^{-1}(1/s) .$

3. (a) $f_1(t) = [u(t)-u(t-1)]\,t \; ; F_1(s) = -\dfrac{d}{ds}\left[\dfrac{1-e^{-s}}{s}\right] = \dfrac{1-e^{-s}-se^{-s}}{s^2} .$

Alternatively: $f_1''(t) = \delta(t)-\delta(t-1)-\delta'(t-1)$

   whence, $s^2F_1(s) = 1-e^{-s}-se^{-s}$;

(b) From (a), $\mathcal{L}\left[\displaystyle\sum_0^\infty f_1(t-n)\right] = \dfrac{1-(s+1)e^{-s}}{s^2(1-e^{-s})}$;

(c) From (a), $\mathcal{L}\left[\displaystyle\sum_0^\infty f_1(t-2n)\right] = \dfrac{1-(s+1)e^{-s}}{s^2(1-e^{-2s})}$;

(d) $F_2(s) = \displaystyle\int_0^1 te^{-st}dt + \int_1^2 e^{-st}dt = \dfrac{1}{s^2} - \dfrac{e^{-s}}{s^2} - \dfrac{e^{-2s}}{s} .$

Alternatively: $f''(t) = \delta(t)-\delta(t-1)-\delta'(t-2)$

   whence, $s^2F(s) = 1-e^{-s}-se^{-2s}$;

(e) From (d), $\mathcal{L}\left[\displaystyle\sum_0^\infty f_2(t-2n)\right] = \dfrac{1-e^{-s}-se^{-2s}}{s^2(1-e^{-2s})}$;

(f) $F_3(s) = \displaystyle\int_0^1 e^{-(s+1)t}dt = [1-e^{-(s+1)}]/(s+1) .$

Alternatively: $f_3'(t) = \delta(t)-\dfrac{1}{e}\delta(t-1)-f_3(t)$

   whence, $sF_3(s) = 1 - \dfrac{1}{e}e^{-s} - F_3(s)$;

(g) From (f), $\mathcal{L}\left[\displaystyle\sum_0^\infty (-1)^n f_3(t-n)\right] = \dfrac{1-e^{-(s+1)}}{(s+1)(1+e^{-s})} .$

CHAPTER 6

Exercises I

1. If $f(t)$ is even then $f(t)\cos n\omega_0 t$ is always even and $f(t)\sin n\omega_0 t$ is always odd; hence

$$B_n = \int_{-T/2}^{T/2} f(t)\sin n\omega_0 t \, dt = 0 \quad \text{for every } n.$$

Similarly, if $f(t)$ is odd then $f(t)\sin n\omega_0 t$ is always even and $f(t)\cos n\omega_0 t$ is always odd.

2. (a) $A_n = 0$ for all n ; $\quad B_n = 2\int_0^{\pi} t \sin nt \, dt = (-1)^{n+1}\dfrac{2\pi}{n}$.

Hence, $f_T(t) = 2\left[\dfrac{\sin t}{1} - \dfrac{\sin 2t}{2} + \dfrac{\sin 3t}{3} - \ldots\right]$;

(b) $B_n = 0$ for all n ; $\quad A_0 = \int_{-\pi}^{\pi} f(t)\,dt = \dfrac{1}{2}[2\pi \times \pi] = \pi^2$.

$$A_n = 2\int_0^{\pi} t \cos nt \, dt = \dfrac{2}{n^2}\{(-1)^n - 1\} \quad \text{for } n \neq 0.$$

Hence, $f_T(t) = \dfrac{\pi}{2} - \dfrac{4}{\pi}\left[\dfrac{\cos t}{1^2} + \dfrac{\cos 3t}{3^2} + \dfrac{\cos 5t}{5^2} + \ldots\right]$;

(c) $f_T(t) = \dfrac{\pi^2}{3} - 4\left[\dfrac{\cos t}{1^2} - \dfrac{\cos 2t}{2^2} + \dfrac{\cos 3t}{3^2} - \ldots\right]$.

The required series in (d), (e), and (f) are obtained by setting $t = \pi/2$ in (a) and $t = 0$ in (b) and (c).

3. $\cos xt$ is even in t so that $B_n = 0$ for all n.

$$A_n = 2\int_0^{\pi} \cos xt \cdot \cos nt \, dt = \dfrac{1}{2}\left[\dfrac{(-1)^n \sin x\pi}{x+n} + \dfrac{(-1)^n \sin x\pi}{x-n}\right].$$

Hence, $\cos xt = \dfrac{2x \cdot \sin x\pi}{\pi}\left[\dfrac{1}{2x^2} - \dfrac{\cos t}{x^2 - 1^2} + \dfrac{\cos 2t}{x^2 - 2^2} - \ldots\right]$

for $-\pi < t < +\pi$. This remains continuous at $\pm\pi$; put $t = \pi$ and divide by $\sin x\pi$.

4. (a) $\cos t = \dfrac{8}{\pi}\left[\dfrac{1}{2^2-1}\sin 2t + \dfrac{2}{4^2-1}\sin 4t + \dfrac{3}{6^2-1}\sin 6t + \ldots\right];$

(b) $\sin t = \dfrac{2}{\pi} - \dfrac{4}{\pi}\left[\dfrac{\cos 2t}{2^2-1} + \dfrac{\cos 4t}{4^2-1} + \dfrac{\cos 6t}{6^2-1} + \ldots\right].$

Exercises II

1. $f_T(t) = \dfrac{kd}{T}\sum\limits_{n=-\infty}^{+\infty}\left[\dfrac{\sin n\omega_0 d/2}{n\omega_0 d/2}\right]e^{in\omega_0 t}$

$= \dfrac{kd}{T}\left[1 + 2\sum\limits_{n=1}^{\infty}\left[\dfrac{\sin n\omega_0 d/2}{n\omega_0 d/2}\right]\cos n\omega_0 t\right],$

equality holding for $nT < t < (n+1)T, n = 0, \pm 1, \pm 2, \ldots$

For $d = T$, $\sin n\omega_0 d/2 = \sin n\pi = 0$; all Fourier coefficients vanish apart from the constant term (consistent with the fact that $f_T(t) = k$ for all t).

For $d = T/2$, $\sin n\omega_0 d/2 = \sin n\pi/2 = \begin{cases}(-1)^{n+1} & \text{if } n \text{ odd},\\ 0 & \text{if } n \text{ even}.\end{cases}$

Hence, $f_T(t) = \dfrac{k}{2}\left[1 + \dfrac{4}{\pi}\left[\cos \omega_0 t - \dfrac{\cos 3\omega_0 t}{3} + \dfrac{\cos 5\omega_0 t}{5} - \ldots\right]\right].$

Finally if $k = 1/d$ and we allow d to tend to zero then $f(t) \to \delta(t)$ and

$\dfrac{\sin n\omega_0 d/2}{n\omega_0 d/2} \to 1.$

In the limit

$$\sum_{n=-\infty}^{+\infty}\delta(t - nT) = \dfrac{1}{T}\sum_{n=-\infty}^{+\infty}e^{in\omega_0 t} = \dfrac{1}{T}\left[1 + 2\sum_{n=1}^{\infty}\cos n\omega_0 t\right].$$

2. (a) $T = 2\pi, \omega_0 = 1$. Hence

$$\sum_{n=-\infty}^{+\infty}\delta(t - 2n\pi) = \dfrac{1}{2\pi}\sum_{n=-\infty}^{+\infty}e^{int} = \dfrac{1}{2\pi}\left[1 + 2\sum_{n=1}^{\infty}\cos nt\right];$$

(b) $\sum\limits_{n=-\infty}^{+\infty}\delta\{t - (2n+1)\pi\} = \sum\limits_{n=-\infty}^{+\infty}\delta\{(t-\pi) - 2n\pi\}$

$= \dfrac{1}{2\pi}\sum\limits_{n=-\infty}^{+\infty}e^{in(t-\pi)} = \dfrac{1}{2\pi}\sum\limits_{n=-\infty}^{+\infty}(-1)^n e^{int} = \dfrac{1}{2\pi}\left[1 + 2\sum\limits_{n=1}^{\infty}(-1)^n \cos nt\right];$

(c) $\displaystyle\sum_{n=-\infty}^{+\infty}(-1)^n \delta(t-n\pi) = \sum_{n=-\infty}^{+\infty}\delta(t-2n\pi) - \sum_{n=-\infty}^{+\infty}\delta\{t-(2n+1)\pi\}$

$\displaystyle = \frac{1}{2\pi}\sum_{n=-\infty}^{+\infty}\{e^{int} - (-1)^n\, e^{int}\} = \frac{2}{\pi}\sum_{n=0}^{\infty}\cos(2n+1)t \ .$

3. (a) If $f(t) = t$ on $(-\pi, \pi)$, assume that $\displaystyle f_T(t) = \frac{1}{2\pi}\sum_{n=-\infty}^{+\infty}C_n e^{int}$

Then $\displaystyle Df_T(t) = \frac{1}{2\pi}\sum_{n=-\infty}^{+\infty}inC_n e^{int}\ .$

But $\displaystyle Df_T(t) = 1 - 2\pi\sum_{n=-\infty}^{+\infty}\delta\{t-(2n+1)\pi\} = 1 - \sum_{n=-\infty}^{+\infty}(-1)^n\, e^{int}$

and so $\displaystyle C_n = \frac{2\pi}{in}(-1)^{n+1}$ for all $n \neq 0$, while $C_0 = 0$;

(b) $\displaystyle D^2 f_T(t) = 2\pi\sum_{n=-\infty}^{+\infty}(-1)^n\,\delta(t-n\pi) = \sum_{n=-\infty}^{+\infty}\{e^{int} - (-1)^n\, e^{int}\}$

whence $\displaystyle C_n = -\frac{2\pi}{n^2}\{e^{int} - (-1)^n\, e^{int}\},\quad \text{for } n \neq 0\ ;$

(c) $\displaystyle D^2 f_T(t) = 2 - 4\pi\sum_{n=-\infty}^{+\infty}\delta\{t-(2n+1)\pi\},$ whence

$C_n = 2(-1)^n/n^2$ for $n \neq 0$.

4. $\displaystyle\sum_{n=-\infty}^{+\infty}\delta\left[t-(2n+1)\frac{\pi}{2}\right] = \sum_{n=-\infty}^{+\infty}\delta\left[\left(t-\frac{\pi}{2}\right)-n\pi\right] = \frac{1}{\pi}\sum_{n=-\infty}^{+\infty}e^{i2n(t-\frac{\pi}{2})}$

$\displaystyle = \frac{1}{\pi}\sum_{n=-\infty}^{+\infty}(-1)^n\, e^{i2nt}\ .$

Also, $\displaystyle D^2\,|\cos t| = 2\sum_{n=-\infty}^{+\infty}\delta\left[t-(2n+1)\frac{\pi}{2}\right] - |\cos t|$

whence $C_n = \dfrac{2(-1)^n}{\pi(1-4n^2)} = \dfrac{2(-1)^{n+1}}{\pi(2n-1)(2n+1)}$.

Exercises III

1. (a) $\displaystyle\int_{-\infty}^{+\infty} f(t-a)e^{-i\omega t}\,dt = \int_{-\infty}^{+\infty} f(\tau)e^{-i\omega(a+\tau)}\,d\tau = e^{-i\omega a}F(i\omega)$

(b) $\displaystyle\int_{-\infty}^{+\infty} \{f(t)e^{-at}\}e^{-i\omega t}\,dt = \int_{-\infty}^{+\infty} f(t)e^{-i(\omega-ia)t}\,dt$;

(c) $\displaystyle\int_{-\infty}^{+\infty} f(at)e^{-i\omega t}\,dt = \int_{-\infty}^{+\infty} f(\tau)e^{-i\omega\tau/a}\,d(\tau/a)$;

(d) $\dfrac{d}{dt}f(t) = \dfrac{d}{dt}\left[\dfrac{1}{2\pi}\displaystyle\int_{-\infty}^{+\infty} e^{i\omega t}F(i\omega)\,d\omega\right] = \dfrac{1}{2\pi}\displaystyle\int_{-\infty}^{+\infty} i\omega\,F(i\omega)e^{i\omega t}\,d\omega$.

2. $h_3(t) = \begin{cases} t+(a+b)/2, & \text{for } -\tfrac{1}{2}(a+b) \leqslant t < \tfrac{1}{2}(b-a) \\ b, & \text{for } \tfrac{1}{2}(b-a) \leqslant t < \tfrac{1}{2}(a-b) \\ (a+b)/2 - t, & \text{for } \tfrac{1}{2}(a-b) \leqslant t < \tfrac{1}{2}(a+b) \end{cases}$

with $h_3(t) = 0$ for all other values of t. (Here we assume $b \geqslant a$).

3. (a) $\mathcal{F}\{e^{-|t|}\} = 2\displaystyle\int_0^\infty e^{-\tau}\cos\omega\tau\,d\tau = 2/(1+\omega^2)$;

(b) $\mathcal{F}\{e^{-|t|}\operatorname{sgn} t\} = -2i\displaystyle\int_0^\infty e^{-\tau}\sin\omega\tau\,d\tau = -2i\omega/(1+\omega^2)$;

(c) $\displaystyle\int_{-\infty}^{+\infty} e^{-t^2/2}\,e^{-i\omega t}\,dt = \int_{-\infty}^{+\infty} \exp\left[-\frac{1}{2}(t^2+2i\omega t)\right]dt$

$\displaystyle = \int_{-\infty}^{+\infty} \exp\left[-\frac{1}{2}(t+i\omega)^2\right]\exp(-\omega^2/2)\,dt$

$\displaystyle = \sqrt{2}\,e^{-\omega^2/2}\int_{-\infty}^{+\infty} \exp\left[-\frac{1}{2}(t+i\omega)^2\right]d\left[(t+i\omega)/\sqrt{2}\right] = \sqrt{2\pi}\,e^{-\omega^2/2}$;

(d) $2 \displaystyle\int_0^1 (1 - t^2) \cos \omega t \ dt = 4 (\sin \omega - \omega \cos \omega)/\omega^3$;

(e) $\displaystyle\int_0^\pi \sin t \, e^{-i\omega t} \, dt = \frac{1}{i\omega} \left[\frac{1}{i\omega} + \frac{e^{-i\omega\pi}}{i\omega} \right] + \frac{1}{\omega^2} \int_0^\pi \sin t \, e^{-i\omega t} \, dt$.

Hence $F_2(i\omega) = \dfrac{1 + e^{-i\omega\pi}}{1 - \omega^2}$;

(f) $F_3(i\omega) = e^{i\omega\pi/2} \, F_2(i\omega) = \dfrac{2 \cos \omega\pi/2}{1 - \omega^2}$.

4. $f(t) = \dfrac{1}{2\pi} \displaystyle\int_{-\infty}^{+\infty} F(i\omega) \, e^{i\omega t} \, d\omega = \dfrac{1}{2\pi} \int_{-\infty}^{+\infty} e^{i\omega t} \, d\omega \int_{-\infty}^{+\infty} e^{-i\omega\tau} f(\tau) d\tau$

$= \dfrac{1}{2\pi} \displaystyle\int_{-\infty}^{+\infty} d\omega \int_{-\infty}^{+\infty} f(\tau) \, e^{i\omega(t-\tau)} \, d\tau = \dfrac{1}{2\pi} \int_{-\infty}^{+\infty} d\omega \int_{-\infty}^{+\infty} f(\tau) \cos \omega \, (t-\tau) \, d\tau$

since $f(t)$ is real. Result follows since $\displaystyle\int_{-\infty}^{+\infty} f(\tau) \cos \omega \, (t - \tau) \, d\tau$ is an even function of ω.

Expanding $\cos \omega(t - \tau)$ gives

$$f(t) = \frac{1}{\pi} \int_0^\infty \{A(\omega) \cos \omega t + B(\omega) \sin \omega t\} \, d\omega \ .$$

If $f(t)$ is even, then $B(\omega) = 0$; if $f(t)$ is odd, then $A(\omega) = 0$.

5. (a) $e^{-|t|}$ even, so that $e^{-|t|} = \dfrac{2}{\pi} \displaystyle\int_0^\infty \cos \omega t \, d\omega \int_0^\infty e^{-\tau} \cos \omega\tau \, d\tau$

$= \dfrac{2}{\pi} \displaystyle\int_0^\infty \dfrac{\cos \omega t}{\omega^2 + 1} \, d\tau$. Hence $\displaystyle\int_0^\infty \dfrac{\cos xt}{x^2 + 1} \, dx = \dfrac{\pi}{2} e^{-t}, \ (t \geqslant 0)$;

(b) For $|t| < 1$, $\quad 1 - t^2 = \dfrac{2}{\pi} \displaystyle\int_0^\infty \cos \omega t \, d\omega \int_0^1 (1 - \tau^2) \cos \omega \tau \, d\tau$

$$= \frac{4}{\pi} \int_0^\infty \cos \omega t \left[\frac{\sin \omega - \omega \cos \omega}{\omega^3} \right] d\omega.$$

Putting $t = \dfrac{1}{2}$ gives $\displaystyle\int_0^\pi \left[\frac{\omega \cos \omega - \sin \omega}{\omega^3} \right] \cos \omega/2 \, d\omega = -\frac{3\pi}{16}.$

Exercises IV

1. (a) $F_1(i\omega) = \mathcal{F}\{u(t)\} - \mathcal{F}\{u(t)e^{-at}\} = \pi\delta(\omega) + \dfrac{1}{i\omega} - \dfrac{1}{a + i\omega}$;

(b) $F_2(i\omega) = \mathcal{F}\left[\dfrac{1}{2}(1 + \cos 2at) \right] = \pi\delta(\omega) + \dfrac{\pi}{2}\{\delta(\omega - 2a) + \delta(\omega + 2a)\}$;

(c) $F_3(i\omega) = \mathcal{F}\left[\dfrac{1}{2}(1 - \cos 2at) \right] = \pi\delta(\omega) - \dfrac{\pi}{2}\{\delta(\omega - 2a) + \delta(\omega + 2a)\}$.

2. $\mathcal{F}\left[\displaystyle\sum_{k=0}^{2N-1} \delta(t - kT) \right] = \displaystyle\sum_{k=0}^{2N-1} e^{-ik\omega T} = \dfrac{1 - \exp(-i\omega T2N)}{1 - \exp(-i\omega T)}$.

$$= \frac{\exp(-i\omega TN)}{\exp(-i\omega T/2)} \left(\frac{\sin N\omega T}{\sin \omega T/2} \right) = \frac{\sin N\omega T}{\sin \omega T/2} \exp\{-i\omega T(N-1)/2\} .$$

Mutiplying this transform by $\exp\{i\omega T(N-1)/2\}$ is equivalent to a displacement to the left of the train of delta functions by an amount $(N-1)T/2$.

That is, $\dfrac{\sin N\omega T}{\sin \omega T/2}$ is the Fourier Transform of $2N$ delta functions disposed symmetically about the origin and spaced T apart.

3. $f(t) \displaystyle\sum_{n=-\infty}^{+\infty} \delta(t - nT) = f(t)\left[\dfrac{1}{T} \displaystyle\sum_{n=-\infty}^{+\infty} e^{in\omega_0 t} \right] = \dfrac{1}{T} \displaystyle\sum_{n=-\infty}^{+\infty} f(t)e^{in\omega_0 t}$

where we write $\omega_0 \equiv 2\pi/T$; now recall that the Fourier Transform of $f(t)e^{in\omega_0 t}$ is $F(i\omega - in\omega_0)$. [This result effectively shows that the inverse form of the Convolution Theorem holds in this case.] On the other hand, the Fourier Transform of $f(nT)\delta(t - nT)$ is $f(nT)e^{-in\omega T}$.

4.   From Question 3: $\mathcal{F}\{f^*(t)\} = \dfrac{1}{T}\left[F(i\omega) + \displaystyle\sum_{n=1}^{\infty} F\left(i\omega \pm \dfrac{i2\pi n}{T}\right)\right]$. If $F(i\omega)$ vanishes

outside $\left(-\dfrac{\pi}{T}, +\dfrac{\pi}{T}\right)$ it follows that the translates $F\left(i\omega \pm \dfrac{i2\pi n}{T}\right)$ do not overlap

and that $H(i\omega)\,\mathcal{F}\{f^*(t)\} = \dfrac{1}{T}F(i\omega)$. Hence the output is $\dfrac{1}{T}f(t)$. Also,

$$h(t) = \frac{1}{2\pi}\int_{-\infty}^{+\infty} H(i\omega)\,e^{i\omega t}\,d\omega = \frac{1}{T}\left[\frac{\sin \pi t/T}{\pi t/T}\right].$$

Thus the output signal is also given by

$$f^*(t) * h(t) = \left[\sum_{n=-\infty}^{+\infty} f(nT)\delta(t - nT)\right] * \frac{1}{T}\left[\frac{\sin \pi t/T}{\pi t/T}\right].$$

Index